How do our bodies fight disease? Written in a clear accessible style, this book gives an up-to-date account of the inner workings of our immune system. Aimed at the general reader, it looks at fascinating areas of science including fever, AIDS and cancer.

Central to the book, from which the title derives, is the idea that in the war against disease our bodies sacrifice millions of cells – some are lost in choosing which cells fight diseases most effectively, others die in these efforts, and yet others die in the recovery and repair processes. The result is a finely tuned and well directed attack against invaders of the body which is aimed at the restoration of good health. Only by pitting all manner of complex immune cells against such infectious agents can we continue to survive in the world.

This book is of interest to anyone who has wondered what is really happening when we become ill and recover.

Marion D. Kendall, Emeritus Professor of London University, is now at the Babraham Institute in Cambridge where her work focuses on the thymus.

Also of interest in popular science

The Thread of Life: the story of genes and genetic engineering
SUSAN ALDRIDGE

Evolving the Mind: on the nature of matter and the origin of consciousness
A. G. CAIRNS-SMITH

Extraterrestrial Intelligence
JEAN HEIDMANN

Hard to Swallow: a brief history of food
RICHARD LACEY

The Clock of Ages: why we age – how we age – winding back the clock
JOHN J. MEDINA

Improving Nature? The science and ethics of engineering
MICHAEL J. REISS AND ROGER STRAUGHAN

Beyond Science: the wider human context
JOHN POLKINGHORNE

Figments of Reality: the evolution of the curious mind
IAN STEWART AND JACK COHEN

Perils of a Restless Planet
ERNEST ZEBROWSKI

Does the Weather Really Matter?
WILLIAM BURROUGHS

MARION KENDALL

Dying to live
how our bodies fight disease

PUBLISHED BY THE PRESS SYNDICATE OF THE UNIVERSITY OF CAMBRIDGE
The Pitt Building, Trumpington Street, Cambridge CB2 1RP, United Kingdom

CAMBRIDGE UNIVERSITY PRESS
The Edinburgh Building, Cambridge CB2 2RU, UK http://www.cup.cam.ac.uk
40 West 20th Street, New York, NY 10011-4211, USA http://www.cup.org
10 Stamford Road, Oakleigh, Melbourne 3166, Australia

First published 1998

Printed in the United Kingdom at the University Press, Cambridge

Typeset in Ehrhardt 11/13pt, in QuarkXpress™ [SE]

A catalogue record for this book is available from the British Library

Library of Congress Cataloguing in Publication data

Kendall, Marion D.
Dying to live : how our bodies fight disease / Marion Kendall.
p. cm.
Includes bibliographical references and index.
ISBN 0 521 58479 5 hb
1. Immune system – Popular works. I. Title.
QR181.7.K46 1998
616.07′9–dc21 98–3646 CIP

ISBN 0 521 58479 5 hardback

Contents

Introduction

Acquired Immunodeficiency Syndrome (AIDS) AIDS patients are severely immunocompromised, since the human immunodeficiency virus (HIV) attacks many of the immune system cells and destroys them. As a result, the patient cannot fight off any pathogens that enter the body through the skin or other major ports of entry, such as the mouth, anus, eyes and sweat glands. Nor can the patient combat uncontrolled growth of cancerous cells within the body. In the initial stages of an HIV infection, before any serious problems are suspected, the skin may have a transient rash. As the disease progresses, patients may develop a number of skin conditions, such as dermatitis around the sebaceous glands of hairs, persistent genital ulcers, and eczema. Fungi such as *Candida albicans*, which can affect many healthy people without harming them, become a problem and cannot be controlled. Later on in the HIV infection, many patients have shingles, and the serious skin and connective tissue cancer, Kaposi's syndrome, takes hold. This is seen as pink/purple blotches on the skin which may ulcerate, and swellings caused by enlarged lymph nodes. As many as 40% of homosexual patients with AIDS develop Kaposi's syndrome, yet in healthy men that are not infected with HIV it is rare and only occurs in older men of certain Mediterranean origins. In Africa, both men and women may develop Kaposi's syndrome in AIDS. Over 70% of patients get infections in the eyes, and these problems are often the first reason for patients seeking help from the medical profession.

Thus, observing the problems that arise in AIDS, and other more commonly encountered diseases that are totally unrelated to AIDS, demonstrates how the normal, healthy body tries to combat disease. The immune system often wins, but if it cannot cope then many conditions can get out of control and become life-threatening.

It was the publicity surrounding the shocking news that millions of people could die from a disease caused by HIV that doctors were unable to cure that made so many people, from a very wide spectrum of backgrounds and lifestyles, stop and think about immunity. The science of immunology is relatively new to medicine compared to other disciplines such as surgery. It is evolving so rapidly that even those involved in immunology every day have to read widely to keep abreast of developments. The more that is known about the processes governing the immune reactions, the more the pathways are found to be astoundingly intricate. Immunologists use a very large vocabulary of specialist words, and often shorten the terms to a series of capital letters; thus, the jargon is impossible for the uninitiated to understand.

This book was conceived to appeal to the very many people who, during their lifetime, have relatives or friends who suffer(ed) from diseases involving the immune system. It is about the modern science of immunology that has grown up around disease and how the body copes with it. Most people will have only a rudimentary idea of how the body functions, so I have started quite simply. As the story unfolds, the descriptions of the various interactions get more complicated, so each chapter can be read on its own and, if the details are not understood, a few paragraphs here and there can be skipped over, and a new aspect of immunology explored with the next chapter (see figure on p. 15).

Immunology is about combating disease. Most people have a knowledge of certain aspects, perhaps of childhood measles, or the concept of vaccination and immunisation in childhood, or before travel. Very few, other than interested specialists, try to grapple with the breadth of the subject. Yet we would all benefit from understanding more about the body's defences to viruses, bacteria and other pathogens in order to cope with life-threatening situations such as cancers, AIDS and autoimmunity. Firstly one needs an understanding of the principles of the processes involved, then to be alerted to the possible interactions with daily life, pollution, diet, etc., before appreciating how immune status affects the quality of life and its outcome. Much of immunology relies upon three tenets: recognition, reaction and recovery. Each of these topics forms a section of this book, and one leads on naturally to the next. Reading the whole book will not allow the diagnosis or treatment of immune diseases, but it may enable a better understanding. I hope that in so doing, the reader will be fascinated to discover the ingenuity of the human body and of pathogens.

The human body is made up of millions of cells. A stereotyped cell was explained to most of us in school, but there is a major step between knowing what a basic building block is, and understanding how cells interact to allow the whole body to function. We all know that an egg can be fertilised, after which it divides and new cells are created that can develop into bone, nerve or muscle, or systems that can allow sights to be memorised and thoughts processed. Furthermore specialised cells can roam the body, recognise invaders in the form of viruses, bacteria or foreign proteins and initiate a chain of reactions that may result in the invader being eliminated. But how do cells recognise each other, interact, pass messages and respond to signals from other cells? Why do cells die, and what effect does the death of one cell have on our own death? How is it that we can live and cope with some harmful interactions from the environment, but others kill or debilitate us?

Such thoughts have puzzled mankind since microscopy revealed cells, so scientists have striven to unravel the mysteries in all ways possible. Today we have an intimate knowledge of cell structure, composition and its genetic potential, and scientists think an ultimate understanding is within our grasp. Even then, we shall only have partial answers to these questions because life processes are not static. With every cell division comes an opportunity to change the building blocks and, in reproduction, genetic recombination allows evolution. This is not unique to Man but also happens to all plants and animals around us.

Major contributions to advancing knowledge in areas that might lead to answering some of these questions have come through the growth in the study of disease through immunology. The progress of this science has enabled giant steps to be made in understanding the life of cells. Immunology is very basic to all life. Not only do the most primitive animals, those without brains or complex sense organs such as eyes, have the ability to mount a protection against foreign (or 'non-self') invaders, but many principles employed are remarkably similar throughout the whole animal kingdom. Such 'conservation', as it is called, is generally recognised as being of utmost importance, as its presence indicates that a basic problem exists which has to be dealt with by many different life forms. It is not surprising then that evolution has allowed the development of many different ways of combating disease, for to fail to overcome infection and illness leads to death or lasting disabilities.

The many ways that are utilised by the cells of the body to control foreign invaders such as bacteria or viruses ultimately result in the death of the invader, or the cells that contain them. Thus the targeted death of cells is a life-giving event. Failure to control invaders means uncontrolled disease and ultimately death of the individual. Because the act of death is so final, and often psychologically painful for those left alive, society often regards death as bad. Inside our bodies, it is rarely destructive to life itself, and is generally the means of saving life. Hence the title of the book – Dying to live.

Acknowledgements

My first thanks are to my husband who is not an immunologist, nor indeed a biologist. He has tried to help me make immunology accessible to more people by pointing out where I was still using too much jargon and needed examples. Old habits die hard, but I have tried. Practically, my husband has taken on numerous day-to-day jobs to enable me to sit at a computer and work on this book. Without his help and support, the book could not have been written. I am greatly indebted to him.

I have also been fortunate in having good friends who have spent time reading and checking various parts of the text. In particular I am indebted to Dr Felicity Nicholson, Dr Pippa de Takats, Dr Ann Clarke, Christopher Trewhella and Tony Moverley for the time and trouble they have spent reading and checking the text. I would also like to thank one of my daughters, Susan Kendall (Rossi), who, despite her clinical commitments, found time for stimulating discussions. If errors persist, it is not their fault, but mine!

PART I – RECOGNITION

1

Molecules of Distinction

Rejecting transplants The techniques involved with modern transplantation surgery are very good today, allowing a wide range of organs to be successfully replaced. However, the patient's own body has unique methods of recognising and acting against bacteria, viruses or many foreign invaders in order to overcome them. The same reactions take place against donor organs since they come from a different body, so powerful drugs are used to reduce these natural, usually beneficial, reactions that are potentially harmful in transplantation surgery. Of great importance is the genetic similarity or dissimilarity between donor and recipient. If a group of genes, known as the histocompatibility genes, are identical in both donor and recipient, then there would be no problems of transplant tissue being rejected. Since all individuals inherit some genes from each parent, the resultant genes in the child will not exactly match those of either parent or any brothers and sisters. So, no two individuals, except identical twins where the fertilised egg was split into two, will have the same set of genes (their genetic make-up). Surprisingly, the immune system's cells can recognise the differences and, as a result, the transplant can be destroyed. In order to ensure success after surgery, the body's own immune system must be dampened down to minimise the chances of any cells reacting to the graft and causing rejection. The patient is therefore very susceptible to infections and diseases during this period, and to cancers in the longer term. Thus, treatments must be finely judged to balance between combating disease and allowing the natural processes to reject the transplant.

How is it that animals know their own species, their mates, and their own young? Whilst sending signals through behaviour is one approach, there are more subtle mechanisms operating on individual cells that allow the uniqueness of the individual to show. The immune system uses these molecules in controlling invaders, and in keeping the whole body healthy. To do this the immune system's cells distinguish 'self' from 'non-self', and so avoid the problems of attacking one's own body in mistake for invading organisms. Key players in immunity are antigens and antibodies. Antigens are signals from bacteria, viruses, fungi, parasites or even plant components such as pollen. Some are recognised immediately, but other signals have to be processed inside our body before the immune system will react to them. The signals are usually given to the immune system in combination with other signs, often from our own cells, to ensure that they are reacted to as required. The antibodies are made by us to fit small parts of the antigen so that the antigen is bound in an antigen-antibody complex. Thus the antigen is inactivated and controlled. Subtle alterations to any of these interactions can cause the reactions to go wrong and result in disease.

This sophisticated use of antibody is a hallmark of a highly evolved immune system that we and other warm-blooded animals have developed over millions of years. Simpler animals use simpler systems. These too need signalling molecules, but they are often very general ones that can be recognised without the need for antibodies. Such systems are quick to respond, and very effective, so we have kept them alongside the antigen/antibody responses, or incorporated them into the whole set of immune reactions that we use.

Today, the greatest need is to exploit these immune reactions of the body to help us control disease. Even when an immune reaction is clearly defined, one finds that the body and the invader can, and do, use alternative methods of achieving the same end. Thus, a new vaccine, for example, may only work in some cases, or against the disease as it exists at that point in time. The result is that drug development is very expensive and rarely, if ever, gives perfect results. The most advanced approaches to drug development now use manipulations of genes. These tiny fragments of protein exist in all living cells and contain millions of instructions for making cellular components; therefore, the instructions are, of necessity, very precise. Changing instructions in genes can now be achieved and, at the end of this book, some ways in which these molecular biology techniques, as they are called, are described. Most of us are afraid of gene manipulation in case irretrievable mistakes are made, but the understanding of genes has gone so far now that we cannot escape the fact that genetic manipulations will be used in future. The more we know about how the immune response works, the more safe and effective new techniques will be.

Similarity and Dissimilarity

One of the extraordinary facts about people is that we are all different, physically and mentally, yet we are sufficiently alike to be able to communicate even when there is no common language. If we need to describe another person, or make an 'identikit', the physical differences between us all may be so slight that it is hard to pinpoint the uniqueness of an individual. Yet our mental, emotional and inner strengths can vary enormously. This is because each person, and indeed every plant or animal, uses the same building blocks, which are called cells, to make their physical form. The result of the construction process depends on the specific plans we have acquired from our parents and also the environment in which we grew up.

The cell itself is amazingly similar in most animals, and shares many of its features with those of plants. However, although built along a basic plan, the outsides of cells can be varied to make a variety shapes and sizes, and the insides adapted to perform different functions. Thus, the human body contains dozens of types of cells. Some make pigment to give the skin its colour, others are able to stretch and contract to make muscles, whilst others are adapted to capture light and allow us to see. All cells, whether they come from the heart, brain, lungs or eye, have a nucleus, which floats in a liquid soup named the cytoplasm, all held together in a bag or plasma membrane which lets nutrients in and waste out in a controlled way (Fig. 1.1). Each cell contains information inherited from both the mother and the father. Since there is so much information concerning the body that has to be passed from the parents to the newly created child, the information is tightly packed together and stored in structures called chromosomes that are contained within the nucleus. Humans have 23 pairs of chromosomes, one set from the mother and one set from the father. Each cell has many pairs of chromosomes and within each chromosome is the blueprint information known as genes. The genes in humans can make about 100,000 proteins! Each cell has a full, complete set of genes although any one cell may not need to use all of the stored information. For example, a nerve cell uses some genes to make the general cell parts, such as the plasma membrane or strengthening structures within the cytoplasm, and in addition will use genes that specify the particular shape and content of nerve cells. Muscle cells may therefore use the same general genes, but not the nerve-cell-associated genes. Instead, muscle cells will use those that create muscle cell shape and function.

The cytoplasm around the nucleus contains all the nutrients needed for a cell to grow and to work, including small 'machines', termed mitochondria, that generate the power for this growth. The molecules within the cytoplasm may be the same in greatly different parts of the body, but how they react together to make the heart pump, the brain think, the lungs breathe or the eye see will depend on which genes in the nucleus are activated. Specific genes can stipulate which products the cell can make: these could be

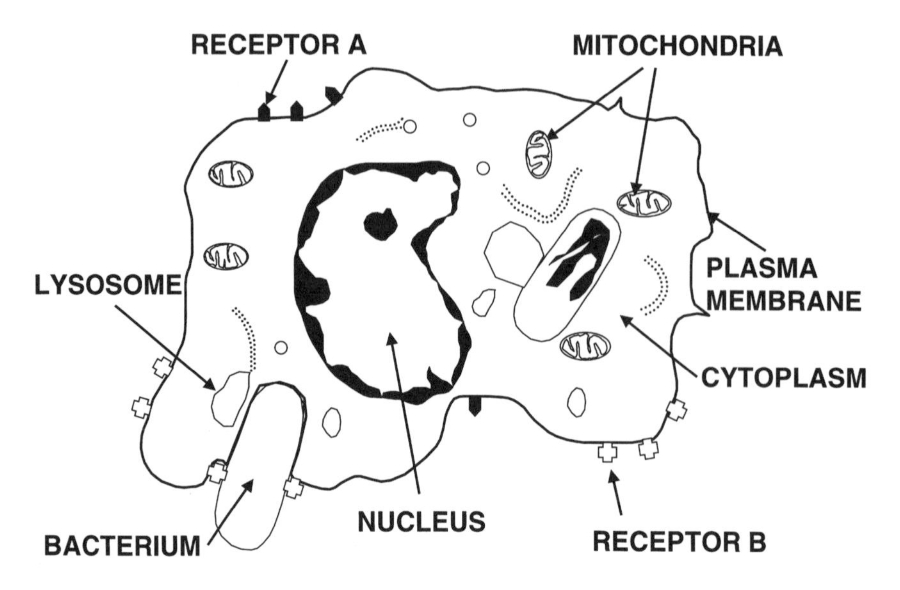

Figure 1.1. A diagram of a macrophage. This cell wanders around in the different organs of the body, where it could encounter any invaders. Special receptors on the surface (A and B are shown here relatively much larger than in reality) help the macrophage to recognise invaders such as bacteria which they can then destroy by taking them into the cell and breaking them up with the aid of bags of enzymes called lysosomes.

pigment for skin colour, contractile muscle proteins, or photoreceptors for the eyes. Ever since living matter was first developed, it seems that the individual cell types have acquired small specialities which make them unique and give them the ability to function in a singular fashion. Thus, the cell is a basic building block that has become adapted to meet the many different demands that life brings.

Most of us know that the body contains a number of vital organs, for example kidneys and the heart, which, if severely damaged, may result in the person dying. The other structures such as bones and muscle can be taken away without causing death. Every part of the body can be recognised by the pathologist from the types of cell present within it. In addition to these characteristically distinct structures, there are two main systems that connect all the organs together, and relay nutrients or messages from one to another. These are the cardiovascular system comprising the heart and blood vessels, and the nervous system which includes the brain, spinal column and individual extensions of nerve cells into each organ. In addition, there are two more systems that utilise the cardiovascular system and the nervous system to send messages throughout the body. These are the endocrine system, which uses small, mainly protein molecules as messengers, and the immune system, which uses different messenger molecules and special cells for its communication network. All four systems, the cardiovascular, nervous, endocrine and immune, pervade the entire body from the outer skin to the innermost organs such as the kidneys, and operate together to ensure that all organs and structures are interconnected so that the body works as a unified whole. The cardiovascular system operates to get oxygen carried by red blood cells (hereafter called erythrocytes) and nutrients to every organ. The nervous system supplies messages by releasing small molecules called neurotransmitters from the ends of nerves to co-ordinate all body functions, and transmits electrical impulses that make muscles work. The endocrine system releases hormones that circulate in the blood and work to regulate and stabilise conditions in the body, even during times of stress or change, such as whilst a woman is pregnant. The immune system acts to keep all of the cells and organs of the body healthy.

Distinctive Markers

How then can the one basic building block, the cell, be altered to achieve such varied roles? One important part of the story is that the cell membrane surrounding and enclosing the cell contains small structures called receptors. These are primarily protein in composition and each cell makes its own set of receptors. Individual receptors come in a variety of forms, are present in large or small numbers, and are arranged to make a cell unique. There are thousands of types of receptor in the body, and a single cell can have 500 to 100,000 copies of any one receptor. Thus, the cells of the heart will have some receptors that are like those on the lungs as well as others that are completely different. These receptors allow messages from the cardiovascular, nervous, endocrine and immune systems, and from other similar cells nearby, to get into individual cells. Attaching a message to a receptor is rather like pressing a key, so that this process can be compared to using a computer. Once all computer programmes are installed, keyboard strokes can open one or more programmes, between which the user can switch. Similar switching-on mechanisms can be used with a variety of types of computer but they will operate different programmes according to how each programme was originally installed. Thus, the assorted receptors on the surface of cells can control events in the cell cytoplasm or activate genes in the nucleus. This means that activation of the same receptor on different cell types can result in completely different actions being carried out.

Each receptor is manufactured by the cell itself. The backbone of the receptor is made of protein, with other molecules, such as sugars, attached to this core. Proteins are mixtures of small molecules called amino acids. Surprisingly there are only 20 common amino acids in living cells, but combining them in different ways creates enormous diversity. The amino acids are linked together to form a secondary structure that is often curled into a coiled helix (the primary structure is the amino acid sequence). The secondary structure is then folded or twisted onto itself to form a tertiary structure or knots of proteins with grooves between them. These grooves are shaped in such a way that only when the messenger makes a 'correct' fit into the space to occupy the receptor (like a key fitting a lock) can the receptor be activated. The molecule that fits into the receptor is called the ligand, and these ligands can be in many forms – from large messages, such as the hormone insulin, neurotransmitters, or chemicals like histamine that is released from special storage cells (mast cells) in the body, to fragments released when cells, or foreign invaders like bacteria, die. This part of the receptor is outside the cell, and it is one of the most important ways in which cells can 'talk' to each other. Another part of the receptor is anchored in the cell membrane and a third, often shorter, part extends into the cell cytoplasm to relay messages to the inside (Fig. 1.2).

When a messenger, or ligand, sits in or occupies its receptor on the outside of the cell, changes occur in the receptor that enable a chain of reactions to be started inside the cell by new messengers. To illustrate this, one can take the computer analogy further. Clicking the mouse button on a computer (arrival of the message) will activate an assortment of programmes or actions that are selected by pointing at an icon. Thus one molecule (for example, the messenger insulin), when it meets its receptor, binds to it very tightly. The tightness of the binding is described as its affinity, and insulin binds with a very high degree of affinity. This makes the receptor carry insulin into the cell where there will be a choice of different pathways that can be activated. Again the computer analogy is useful. Once a programme such as a word processing package is open, keystrokes or mouse pointers can perform many actions that give totally different results. Often the programme contains predetermined pathways already saved as 'macros'. Once activated, a series of steps happen, apparently automatically, to achieve an end. The cell got there first. Inside every cell, there are dozens of 'macros'

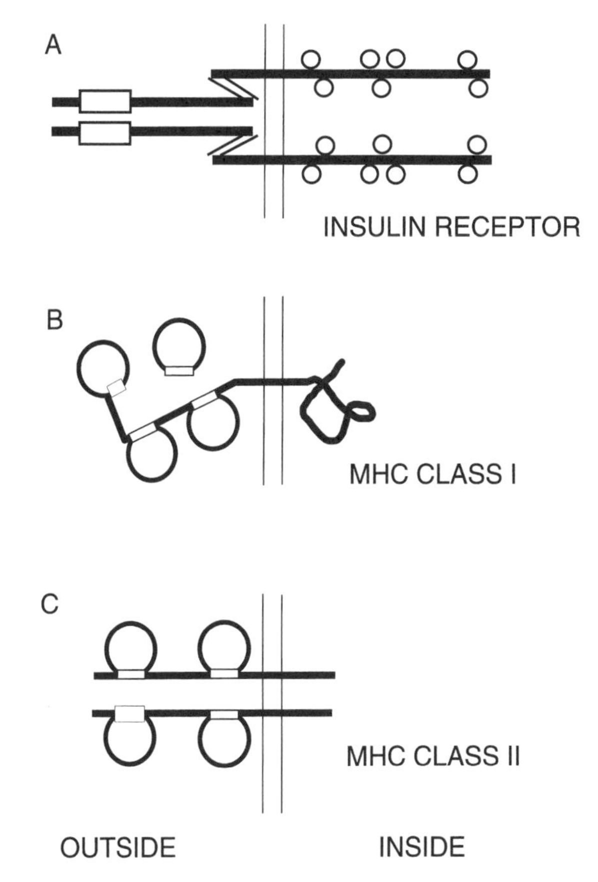

Figure 1.2A–C. Three type of receptor found on cells. A. The insulin receptor binds insulin on each of its component chains (shown as two clear oblong boxes), and they communicate across the cell membrane to activate regions (clear round circles) which send out new messages causing the cell to take up sugars and turn them into the storage form, glycogen. B. The class I major histocompatibility complex (MHC) receptor also spans the membrane and has three regions of globular proteins (incomplete circles stabilised by sulphurous bonds shown as narrow clear oblongs) on a long chain. This receptor interacts with T cells of the CD8 subtype in immune reactions. C. The MHC class II receptor is slightly different and this will interact with CD4 T helper cells.

and they are dependent on what the biologists call 'second messengers'. These substances are key molecules that, once activated, start a succession of steps inside the cell. There are a limited number of 'second messengers' within the cell, not a unique one for each messenger molecule. Indeed each second messenger can be activated by many different messengers. Some drugs work by mimicking or stopping the action of naturally occurring message molecules.

The steps initiated by second messengers are generally very complicated. In most cases they result in amplification of the original binding between the receptor and ligand by their own new interactions and also set off other second or tertiary messengers to undertake a variety of different functions. Specifically, this might be the manufacture of a protein. In order to do this, a gene in the nucleus of the cell will have to be read to extract the recipe for the ingredients, and messages transported to the cytoplasm where the basic components of amino acids, energy, and possibly sugars and fats are organised to be ready for use. Obtaining the energy is often provided by splitting molecules and releasing the power that was used to hold them together. Again the cell uses macros, so that making a protein may use several smaller macros built into larger macros. In another example, the single action of binding insulin to the cell surface will result in the cell taking up the sugar, glucose, from the blood, and turning it into another sugar form, called glycogen, that can be stored in the cell for future use. At the same time the cell is prevented from breaking down any stored glycogen, or other storage molecules such as fat. There are many other consequences of ligands binding to receptors; for example, pores in the membrane may be opened to create channels that let nutrients in, a cell may be stimulated to divide, or the release of other messengers from the cell can be prevented, controlled or started.

In order to stop accidental messages getting through, some receptors act in dissimilar pairs or groups of receptor complexes, so that two or more unique molecules have to fit into closely associated individual receptors before the cell reacts. After a receptor has been occupied by a messenger, the messenger and the receptor are often taken into the cell, the messenger broken down and the receptor recycled out onto the surface again for reuse. During this time there may be fewer receptors on the cell's surface, so the cell cannot be stimulated as much as before. But if more receptors are needed, the cell can make more. A good example of this is the receptor for cholesterol, which is the basic component for making the steroid sex hormones such as oestrogen and testosterone. If there is more cholesterol in the blood, then the cells add more receptors onto their surfaces, and more cholesterol is absorbed. The number of cholesterol receptors on cells also increases with increasing age, but no-one knows why.

Distinguishing People

In the way that invaders generally have distinctive markers on or in them, so our own cells each have a marker on them signalling that they belong to the same body. These markers, made by us, are unique to each person, except for identical twins who have identical markers. The body uses them to distinguish self from non-self, and this fact emerged clearly in transplantation surgery when tissues from donors did not grow and were rejected. The more closely related two people are, the better the chances of successful transplantation surgery. Thus these markers were given the name of histocompatibility antigens, indicating that these antigens allow the mutual acceptance of tissues ('histos' is Greek for tissues).

Each cell makes its own histocompatibility antigens, and puts them on the cell surface where they can be recognised by receptors on other cells. Which of the histocompatibility antigens are present on a person's cells is controlled by a complex of several hundred genes known as H or histocompatibility genes that are responsible for specifying which proteins are used as recognition markers. Some of these proteins signal male or female, and others label the cells as coming from our own body. The largest and most studied group of histocompatibility genes is called the major histocompatibility complex (or MHC) which has been extensively studied in mice. In humans, the MHC genes are situated on one chromosome number 6, and the human antigens produced by this complex are called human leukocyte antigens (HLA). As with all pairs of chromosomes, one is inherited from the mother and the other from the father, so that no two individuals, except identical twins, will have the same genetic composition.

Histocompatibility antigens also affect how we fight disease. Most of us know that different people belong to different blood groups, and recognise the importance of matching blood types in transfusions. Blood groups are defined by the collection of antigens on the surface of each erythrocyte (red blood cell). All erythrocytes in one individual person have the same antigens. The concepts for tissue transplantation are the same, but there are many histocompatibility genes, and therefore many surface antigens. These are classified into three groups: class I, II and III. Class I antigens are on the surface of all cells of the body that have a nucleus, whereas class II antigens are only on certain cells. The most important are white blood cells that are either B cells which produce antibodies, macrophages which remove pus and other material from a site of infection, and T cells that are being used actively in an immune reaction. The major histocompatibility genes are the most studied, but there are also a large number of other groups of genes which control which minor histocompatibility antigens are made. Some of these cause organ rejection a long time after the patient thinks that the operation has been successful. However, the closer the MHC matching, the better the chances of transplants being accepted.

How may histocompatibility antigens affect disease? It seems that if a person is born with certain histocompatibility genes, then this makes them more likely to get specific diseases later in life that result from the immune system attacking the body's own cells. These are therefore called autoimmune diseases. These diseases, which include such conditions as rheumatoid arthritis, multiple sclerosis and several diseases of the thyroid gland, will be considered in more detail in Chapter 9. Most of the diseases are not related to each other but, because cell recognition goes wrong, one person may develop several diseases at once. The genetic component means that autoimmune diseases often occur in families and can be more common in some populations of people than in others. As just described, the MHC genes are mainly on chromosome 6 and clustered into locations, or loci, on the chromosome named alphabetically. These are further subdivided to specify individual gene groups. Thus it is possible to identify which group of genes is associated with which diseases. People with MHC genes at specific points in the D group (DR3/4) and B group have a greater than average relative risk of developing Addison's disease, in which the function of the adrenal gland is altered, type 1 insulin-dependent diabetes and/or rheumatoid arthritis. Indeed the DR3 locus has been called 'the disease susceptibility locus'.

However, it is the total spectrum of MHC genes held by an individual that is important in protection against or susceptibility to certain autoimmune diseases. The incidence of the HLA-DR3/DR4 molecules is higher in Chinese and Japanese patients with insulin-dependent diabetes mellitus, than in those without the disease within the same populations of peoples. If these diabetic patients also have certain other MHC genes, called DQ, they are even more likely to be susceptible to this form of diabetes, although the same combination of DQ genes in other ethnic groups where the DR groups are different does not confer susceptibility. Thus the same human leukocyte antigen (HLA) molecules can be protective in some populations and yet render other people more at risk. As the best way to anticipate a long life is to have parents that were long-lived, so people with family histories without any autoimmune diseases may also have a good chance of not developing these conditions in their own lives. But even with the 'risk-associated' genes, an individual receives some protection from the presence of other MHC gene combinations, such that no autoimmune diseases may develop.

Fascinatingly, there are other ways in which the histocompatibility antigens may be influencing our lives. The first has really only been studied in rodents such as rats and mice. It seems that these unique MHC antigens can be recognised as an odour and this is used as part of the behavioural responses whereby rodents mark their territories with their urine. It has been shown that the immune and blood-forming cells of the bone marrow are those cells responsible for the MHC 'smell', and that if bone marrow transplants are given to mice, their smell changes. It has long been known that odours are specific to individual animals of many types, and that they change, even in humans during some illnesses. In insects, the term pheromone has been coined to describe some chemical signals that are only needed in the minutest of doses, perhaps just a few molecules, to attract mates. Whether MHC antigens and pheromones are part of the same signalling methods, and what humans have in this respect, is not yet clear. We do know though, that people can, without realising it, in some, as yet undefined manner, be aware of major histocompatibility antigen differences between people. It is said that couples who are attracted to each other often have disparate MHC groupings and, furthermore, that if the histocompatibility antigens are very similar then such couples may have problems in conceiving children. Amongst women suffering spontaneous abortions, there is a high rate of couples with similar MHC genes. Other problems with conceiving after *in vitro* fertilisation appear to be linked to the presence of other MHC genes, but the whole subject of MHC genes and fertility needs far more study before it is fully understood.

Our own ability to 'recognise' the major histocompatibility complement of another may also be important in stopping incestuous relationships. It appears to be an advantage to introduce new genes into a population, as shown when in-breeding occurs in domesticated animals and resistance to disease results. It is just as biologically advantageous for Man not to interbreed. Inter-breeding restricts the number of genes that can recombine to create new offspring or, in the case of MHC genes, limits the number of new cells with different antigenic recognition capabilities. The greater the potential to create variation, the greater the chances of developing a wide range of cells able to participate in antigen recognition, and the better the chances of dealing with viruses, bacteria and other pathogens.

Thus not only do people differ from each other in facial and physical characteristics, but we also have an enormous number of subtle molecular differences inside us, of which we are not generally aware. The 'instinctive' attraction between people is fundamentally more important that we realise, and when this leads to the introduction of new genes into a population it is almost always to the advantage of the population in terms of improved health of the offspring.

Identifying the Enemy

In the above section, we have just seen how most cells of the body have a unique form of MHC class I antigen on their surfaces that classify all the cells as belonging to one person. The body's own immune system can read this signal, and use it in the fight against invaders. Such invaders – which can be a whole range of lower animals from single-celled organisms, such as bacteria, to small, complex animals like parasitic worms – also carry their own form of characteristic signals. These molecules are not like MHC class I signals, since they are not usually characteristic of one individual, but they are molecules that most members of a family or species within a family carry on them. The special feature of our immune system's cells, unlike any other system in the body, is that they can make receptors, either of one kind or groups containing dissimilar receptors, that can recognise these foreign antigens. To ensure that such foreign molecules are not confused with antigens derived from our own body's cells, the immune system's specialist cells do not just rely on recognising the invader's antigen, but they are only activated when the foreign antigen is recognised alongside the MHC molecules discussed above. In order to complex the invader's antigen with the body's own MHC antigens, the invaders are taken into cells, and the foreign antigens are cut up to produce small fragments that can be associated with newly formed MHC inside the cell. The new complex can finally placed on the cell surface (Fig. 1.3). Such precise antigen-processing actions are carried out by highly specialised cells called antigen-presenting cells. These cells therefore ensure that at least two signals are present before an immune cell is activated. At this stage, other cells of the immune system can 'see' the antigen, recognise it as foreign, and try to destroy it. However, for this type of immune reaction to take place, the immune cells also need yet further signals before they are active. Such double signals reduce accidents whereby the body's own cells might be destroyed. Sadly, such adverse reactions do occur in the autoimmune diseases, when the body is actively attacked by its

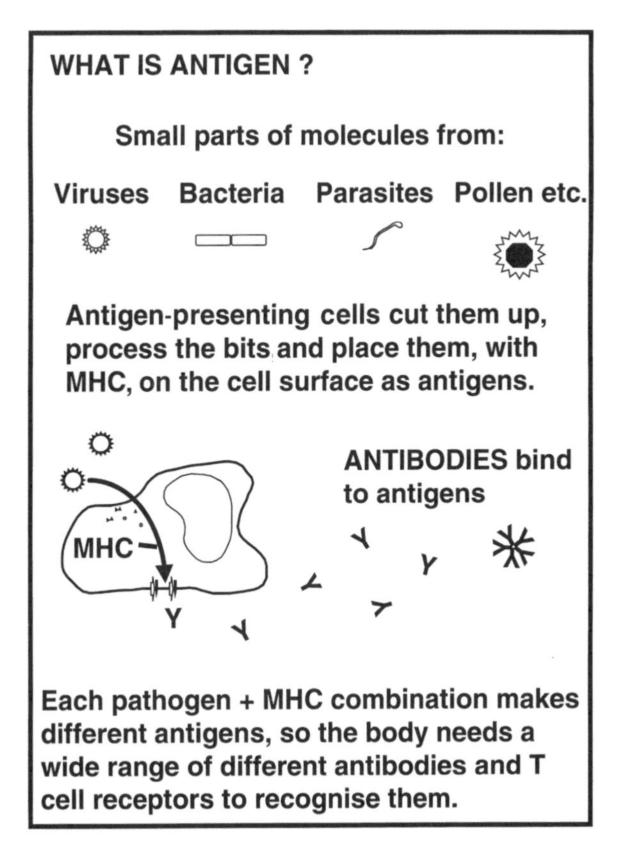

Figure 1.3. Antigen can be recognised in its own right, or when processed through antigen-presenting cells as shown here. MHC proteins are then added to the newly cut antigenic molecule and placed on the cell surface where antibodies (or T cells, not shown) can recognise them.

own immune system. We still do not understand what has gone wrong with the recognition processes in many of these diseases, but they will be discussed more fully in Chapters 7, 9 and 10.

Many of the most important immune system receptors for antigens share certain structural relationships, and are therefore classified as members of one large family – the immunoglobulin superfamily which will be described in more detail in Chapter 5. This superfamily is the largest family of recognition receptors in the body. Most of these receptors occur on immune system cells, but some also occur on cells of the nervous system, especially in the brain. Immunoglobulins are structured so that each part has a specific function. The ends of the immunoglobulin have arms on them that can wave about and form recognition sites where antigen is bound. The nature of the binding site can be varied by unique genetic processes described in Chapter 5. The effectiveness of this variation in structure determines how we cope with antigens from the many different types of invaders such as viruses, bacteria, fungi and parasites. These variable parts are on the arms and they are attached to a non-variable region of the receptor and then to a stem. Some immunoglobulins are attached by the stem to cells, but many are released into the blood from the cells that make them. They can then directly combine with antigen and mop it up.

The complexities of the antibody and antigen interaction form the Humoral Immune Response (see figure opposite). This is one part of an adaptive immune system which more highly evolved vertebrates such as humans, other mammals and some cold-blooded animals have perfected to cope with a wide variety of potentially harmful pathogens. The other part is called the Cell-mediated Immune Response because it destroys pathogens by killing the infected cells. A key feature of both adaptive immune responses is the precise recognition of antigen. Once recognised and reacted against, the adaptive immune system creates a memory for the antigen, so a second invasion can be reacted to quite quickly.

A more basic form of immunity, used by a much wider range of animals including many single-celled or cold-blooded animals, is the Innate Immune Response. This is simpler, and relies primarily on the ability of certain cells, called phagocytes, to swallow pathogens and kill infected cells. Although the innate immune system also relies on receptors, they are not as specific as those used by the cells of the adaptive immune system. Furthermore they are not capable of being varied in structure, like the immunoglobulins described above. The non-specific nature of the receptors means that there is no memory aspect of this type of response. Thus the innate system provides a quick first line of defence response to invaders, but is less specific than the adaptive immune response.

The division into innate and adaptive immunity is useful, but we now realise that there is no sharp distinction between the two systems, as factors released by one type of response can trigger reactions of the other type of response. Indeed, all immune responses use complex cell signalling to recruit other cells into the control processes, and the chain of events is diverse and intricate. It is also fascinating. It is hard to believe that such complicated systems can have been evolved without us knowing anything about it. Modern Man, even with the most powerful computer, would find it

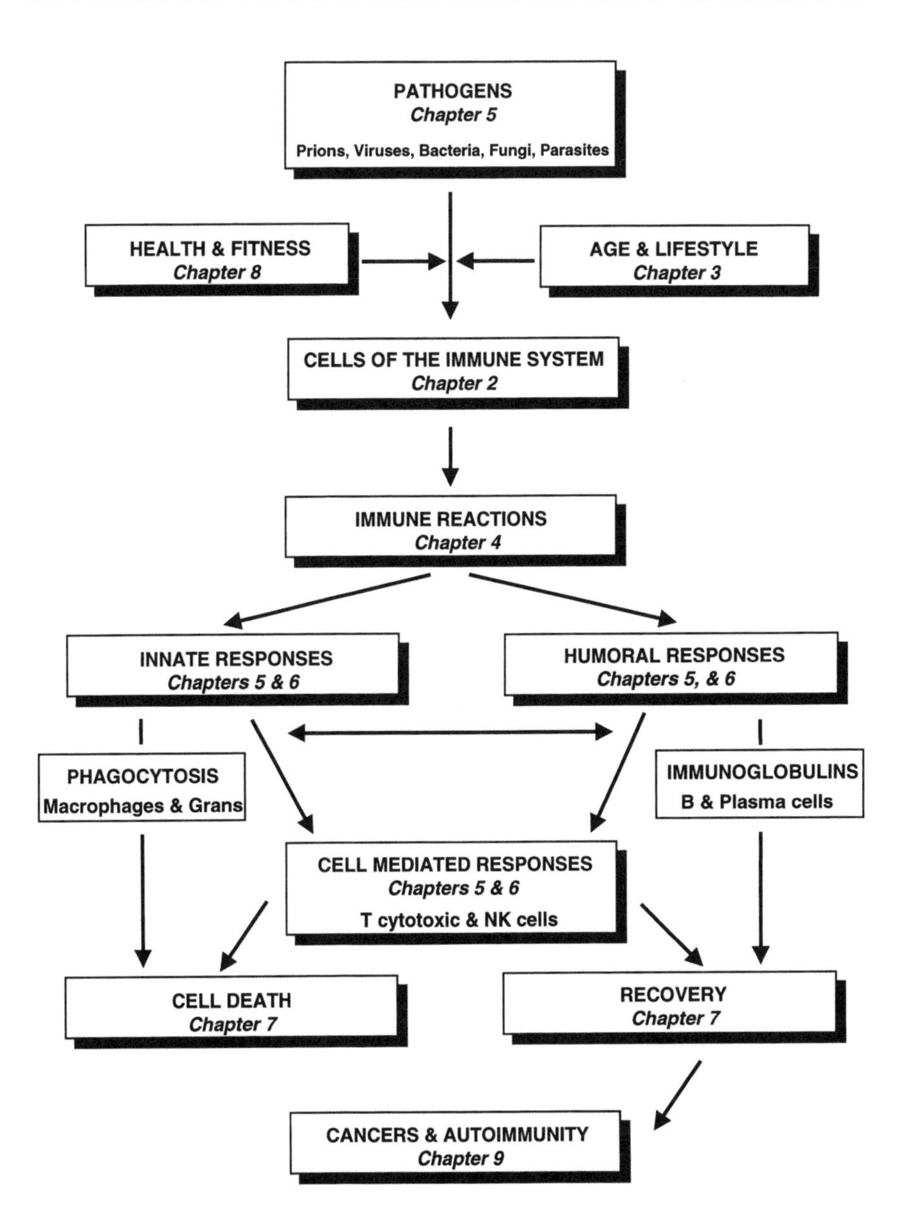

difficult to think up even some of the myriad ways that pathogens are controlled in our body. Nature often seems to choose several approaches to the same problem, so that if one does not work, another may. It is not surprising that it is difficult to design drugs to interfere with one system without upsetting many other control circuits. Hopefully the complexities of the systems will not deter the reader from the fascination of immunology that follows, as the future for immunology is exciting. Amazingly, research has identified key molecules that can be artificially used to work wonders in controlling disease. It is a real prospect that soon the location of all our genes will be known, so that we can envisage that the nature of all gene products will be determined, thus enabling new disease therapies to be designed.

2

Distinctive Families

Leukaemias These are the most common cause of malignant disease in children. However, leukaemia can develop at any age and, in most countries, the majority of people who get leukaemia are over the age of 55. The commonest childhood leukaemia is acute lymphocytic leukaemia whereas aging adults tend to get chronic granulocytic leukaemia. There are multiple factors that cause leukaemias, but there are some groups of people who are more likely to get them. These include people exposed to ionising radiations or to certain chemicals, or those born with particular congenital abnormalities. Characteristically, there is a problem in the production of new blood cells in the bone marrow of the body. The most primitive blood cells in the series develop faults, and they cannot be induced to change normally into mature forms which should circulate in the blood. These immature, malfunctioning cells accumulate in the bone marrow, and upset normal blood cell development. Thus the blood may contain fewer cells than normal of a particular type, and in addition a few leukaemic cells may escape into the blood. So an examination of the blood in patients reporting aching and tender bones, enlarged joints or severe anaemia is generally followed by an examination of the bone marrow to see if abnormal cells are present there. Often in acute leukaemia the lymph nodes and spleen swell. Treatment of leukaemias today is quite effective, but death may result from secondary problems such as infections, bleeding and failure of the kidneys and/or liver. However, patients can be cured, and more leukaemic patients live longer than ever before. Finally our increased awareness of environmental causes has led to reduced risks for many people.

The Family Tree

Cells, like people, belong to families. There is a family of muscle cells, mainly in the limbs, that contract only when told to, and other muscle cells that contract involuntarily, without us knowing anything about it. These involuntary muscle cells are in the heart and other organs inside us. When we eat a meal, the food must be moved along the intestines by muscles that work all the time to contract and relax: we are not able to consciously decide when these muscles will contract. There are numerous families of cells within the body, each derived from very unspecialised stem cells which have the potential to develop into other more specialised cell types. When stem cells divide they can develop along new lines that allow them to give rise to different types of cell. For example, the cells that make up bone all derive from stem cells called precursor osteoblasts. These cannot make bone yet, but as they develop and divide they start to secrete bone around them and then they become totally specialised to handle the calcium and phosphate minerals needed to form mature bone. The fully specialised cells are called osteocytes, and these can no longer take on any other functions in the body, other than to make bone. Each organ contains cells characteristic of the organ, and are often so specialised that they can only function for that organ.

Just like the heart, liver and other organs of the body, the organs of the immune system are also made of specialised cells, and their very important secretions. Immune system cells are produced in the bone marrow or a gland over the heart called the thymus, from where they enter the bloodstream. When these cells meet and recognise pathogens or foreign antigens they will be stimulated to make an immune response. The full immune response can take place anywhere in the body, but generally it is within special lymphoid organs such as the lymph nodes, spleen, or tonsils. Then cells or their secretions move away from these sites to go all over the body. All of these immune cells belong to the family of white blood cells which are technically called leukocytes. Red blood cells, or erythrocytes, give blood its colour and carry oxygen to every organ. There are many more erythrocytes than leukocytes in blood, but the reverse is true in the body's organs since leukocytes can move out of the blood but erythrocytes are restrained within blood vessels and are therefore rapidly carried through the organs. In a litre of blood, an adult man may have 51,000,000,000,000 erythrocytes and 72,000,000,000 leukocytes or 1 leukocyte to about every 700 erythrocytes. Women generally have slightly fewer erythrocytes and very slightly more leukocytes than men.

The blood contains a wide range of different mature leukocytes (Fig. 2.1), but their more primitive forms are contained within the bone marrow or lymphoid organs. Most of the leukocytes in the blood are lymphocytes. These have a simple rounded nucleus and very little cytoplasm. Others, granulocytes, have a nucleus that is lobed and twisted like a string of beads, and have more cytoplasm that is packed full of granules. Each of these two major families are composed of subfamilies of cells which share some characteristics, yet also have specialisations to undertake specific roles in

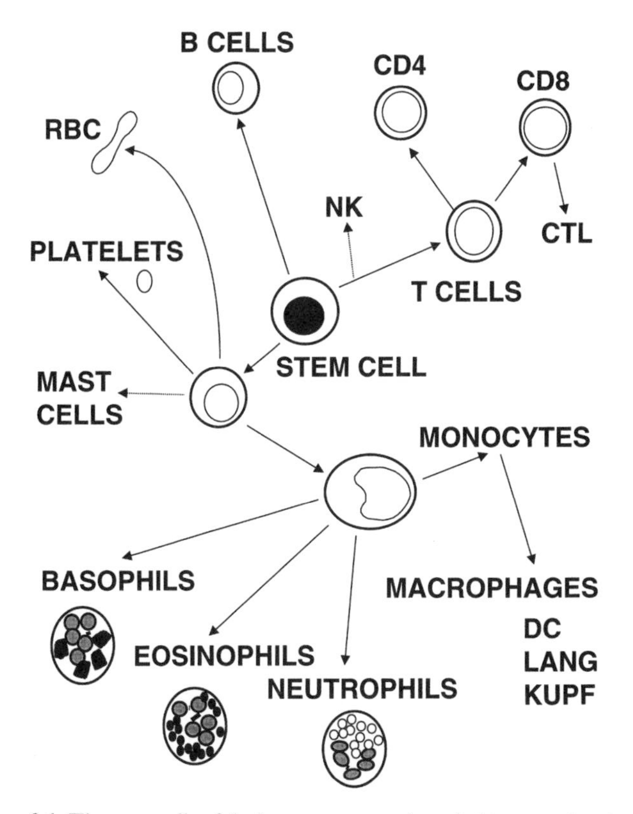

Figure 2.1. The stem cells of the bone marrow or lymphoid organs develop into a variety of leukocytes. The B and T line of lymphocytes includes NK (natural killer) cells and CTL (cytotoxic lymphocytes). The macrophage line of DC (dendritic cells), Lang (Langerhans's cells) and KUPF (Kupffer cells) all derive from monocytes. RBC is the abbreviation for red blood cells (erythrocytes).

keeping the body healthy. The blood also contains smaller numbers of other cells such as the precursors of macrophages known as monocytes, or a very, very small number of stem cells. Neither of these types of cell is functional in the blood, but both use the vascular system as a highway to reach different organs where they can develop further. Another component of blood is numerous small fragments of cells, without a nucleus, that are called platelets. These are very important in forming clots to prevent blood leaking out of damaged blood vessels, and they contain a wide variety of chemicals many of which speed up repair processes.

Most lymphocytes are either T or B cells, but small numbers of other cell types such as natural killer cells also exist. T is short for thymus, which is where these T cells are made, whereas B cells are primarily made in the bone marrow. During their development both T and B cells acquire unique receptors on their cell surface which are used to distinguish foreign antigen from the antigens of one's own body. Any cells that might potentially harm the body are killed off in the thymus or bone marrow before they can get out into the blood. However, in certain disease conditions the blood may contain some harmful auto-reactive cells, but this is not the normal healthy situation.

Family Estates

Most cells need oxygen to enable them to grow. In the very first days, as the baby forms in the womb which is more correctly termed the uterus, oxygen just diffuses across cell membranes, but as the cell masses become larger and more complex there is a need for erythrocytes to carry oxygen into the tiny embryo. The very first blood-forming cells are called haemopoietic cells from the Greek word 'hemo' for blood, and 'poiesis' indicating formation. These haemopoietic cells form islands of minute groups inside the embryo, and then some move outside into the membranes around the embryo where there is more space for them to multiply. When the embryo is about two months old, some cells become special large erythrocytes that move back into the embryo. This coincides with the formation of very tiny blood vessels, and a primitive circulation starts to be present. The new blood cells move around in these blood vessels and take oxygen to each developing organ. As this occurs so the primitive cells of the immune system start to collect in places in the embryo where they have the correct microenvironment for development.

In these early first few weeks of growth of the embryo, there is not yet any bone marrow because the bones are not yet formed. Neither has the thymus or spleen developed yet. However, the embryo needs a liver from the earliest possible time, as the liver provides energy and detoxifies waste, so this is a large organ in the embryo. The primitive stem cells capable of making the immune system find the liver and make this their first home where they can divide, differentiate and form the cell lines described in the previous section. The liver then is the first family estate in the embryo.

Fairly soon, when the embryo is no more than 5.5 millimetres long, or about four to five weeks old, some primitive embryonic blood cells move out of the liver and travel around the body. But it is not until the baby is about eight to nine weeks old that these early cells can enter the first blood-forming organs. The first to form is the thymus for T cell development, and the spleen where all types of immune cells develop. The thymus is the major site of production of all of the T cells of the body, and it is particularly active in the embryo and in children. Similarly, the spleen and the liver make all blood cell types in the developing baby, but their ability to do this is reduced as pregnancy proceeds and is almost negligible when the baby is born. The liver then takes over other non-haemopoietic functions, and the spleen both controls the number of erythrocytes in the body and makes leukocytes.

Although the bones start to form quite early in development they are made of soft and bendable cartilage for a long time in the uterus. As the baby grows, bone forms around the cartilage and gradually replaces most of it. With the young cartilaginous bone supported on the outside by a ring of bone, the central region can be used for other purposes. The earliest bone marrow development is seen from about seven to eight weeks from conception, and blood cells are formed from about ten or eleven weeks onwards in most embryonic bones including the long bones of both limbs. By the time the baby is born all of its bones have bone marrow inside them. This can be seen in a newborn baby if their hands are seen against a bright light. The fingers all appear red from the blood and bone marrow in every bone. This special environment of bone marrow has all of the growth factors to allow all blood cells to be made. However, the main families of cells made in the bone marrow are the erythrocytes, lymphocytes of the B cell and macrophage lineages, and white blood cells called granular leukocytes.

Neither the newly produced T cells, nor the B cells are mature at this stage. Although they have the receptors to interact with antigen, they are not functionally mature until they meet antigen. All this time the baby is protected from the outside world, and therefore most foreign antigens, by being in the womb. As soon as the baby is born however, it is exposed to all of the pathogens of the world, and so it is more easily infected. Immune reactions need special organs where immune system cells can expand in numbers to fight any infection whilst not interfering or interacting with virgin T and B cell development. These lymphoid organs are called secondary lymphoid organs, and are described as antigen-driven lymphoid organs. Most secondary lymphoid organs are situated close to where infections start; for instance around the throat, in the walls of the gut and lungs, and all over the body near vital organs. The tonsils and adenoids are around the throat and nose, the gut regions have lymphoid tissues associated with the moist membranes, and the general organs all over the body are the lymph nodes. The spleen is also an important secondary lymphoid organ that also checks the blood for pathogens.

Almost everyone has suffered swollen tonsils as a child, and knows how they can be enlarged and become painful. The tonsils are at the back of the throat and, in addition, the soft skin in the throat and at the back of the tongue has accumulations of immune cells which can react to incoming pathogens. This accumulation of lymphoid tissue is called Waldeyer's Ring. It completely surrounds the entry points for pathogens from the mouth and nose. The adenoids are close by, up in the airway between the nose and the back of the throat. Normally in children, the tonsils are in a resting state and fairly small, but with an infection all of the throat becomes sore as they enlarge. Inside the tonsils, spherical regions called germinal centres are formed and it is here that B cells are made. These B cells of are two main types, those that can produce secretions of antibody to react with the infection, and cells with a memory for the infective agent. This swelling is a response to a specific antigen. So, swollen tonsils are reacting only to the bacterium or other pathogen that got into the throat or nasal passages. Similar reactions to antigens occur in lymph nodes, adenoids and in the tissues associated with the moist linings of the gut and respiratory systems.

Each of the major organs and regions of the body has a collection of lymph nodes situated where they can react to the entry of pathogens. Swollen lymph nodes therefore indicate to a doctor where there is an infection. For example, swollen lymph nodes in the groin will occur when the urinary or genital tract, or legs have an infection. Lymph nodes are connected to one another and to the blood stream by lymphatic vessels. These form a network throughout the body that can drain fluid from an infected region into the nearest lymph node where an immune reaction causes them to swell. Smaller and less organised lymphoid aggregations occur where each of the airways branches in the lungs, and along the walls of part of the gut. We are not usually very aware of changes or swellings in these, but an extreme example is the appendix when it becomes enlarged. In addition to making new cells in response to pathogens in the gut, the appendix traps some of the food that is in the gut and this cannot get out. The whole area becomes inflamed and the appendix can eventually burst. This is quite dangerous so the patient has to be rushed into hospital to have all of the infected fluid drained away as soon as possible, and the appendix can be removed to prevent similar re-occurrences.

The spleen, which is situated under the ribs on the left-hand side of the body, is a large and complex organ, since it serves two main functions which are completely different. It is organised so that all of the blood of the body is filtered through it and this means that erythrocytes (red blood cells) can be inspected to see if they are healthy. Erythrocytes are unusual cells since they do not have a nucleus and are therefore destined to die after about 120 days. As they age they become more rigid, and this change in flexibility can be recognised by cells in the spleen, so that these old erythrocytes can be taken out of the blood and broken down in a part of the spleen called the red pulp. Regions of red pulp are separated by white pulp. The white pulp is a secondary lymphoid organ, specialised to react to any leukocytes (white blood cells) from the blood carrying infective agents. If any antigenic material is found, an immune reaction is started by activating other cells in the spleen to mount an immune response against these foreign invaders (Chapters 5 and 6).

The T Cell Family

In order to recognise any of the large number of antigens that may get into the body, we need many T and B cells, each with a different receptor on their surface. Thus, hopefully, one or two cells out of the whole, wide repertoire of T or B cells can interact with the antigen. This degree in variation of the cell's characteristic receptors is achieved during development using a unique method of gene rearrangements to be described later in this section. No other cells of the body use this mechanism of gene rearrangement, and thus no other cells have a diversity of receptors like those of T and B cells.

The fully developed T cell receptor (Fig. 2.2), like most other receptors, has a long region that is outside the cell, and some shorter regions inside that interact to send messages to the cell when the receptor meets antigen. The really amazing parts are the outermost regions, which can be varied during development to produce unique T cells. This variation is produced in the thymus, and there unwanted developing T cells are not allowed to mature or leave the thymus. Only T cells that cannot do damage to one's own body survive. Developing T cells are said to be educated and selected in the thymus. No other organ appears to have the ability to undertake this job.

Immunologists have spent a lot of time trying to work out the details of the development of the T cell and its receptors, but the way that such a wide diversity in T cell receptor for antigen is achieved is complicated. Essentially the genes that determine the variable end part of the T cell receptor are unusual in that they do not lie together on the chromosome. This means that there is unrequired genetic DNA on the chromosomes. This can be cut out in many different ways by a process called gene rearrangement. Depending on how this is done, the genes that remain come together to make different blueprints for the molecules to be included in the receptor. Once the genes have been 'read', a set of instructions for making the receptor is carried out of the nucleus by messenger ribonucleic acids (mRNAs) into the cytoplasm where the molecules are to be made. This mRNA can also be varied in the cytoplasm in these or other cells, so that there is enormous potential for varying the structure of the T cell receptor. Since there are about 100 genes controlling the variable part of the receptor, numerous joining genes, and multiple combinations of mRNA arrangements, the end result is that the glycoproteins produced are immensely variable. It has been estimated that the recombinations available could be 100,000,000,000. The chances of guessing the winning six-figure number combination in the UK National Lottery are much better than the chances of anticipating which kind of T cell receptor an individual thymocyte will make!

This presence of the T cell receptor distinguishes T cells from other immune system cells. The cells use it in conjunction with another T cell receptor called the CD3 receptor. They act together as a complex to bind antigen. We have already heard much about MHC. This is where it really becomes important in immune responses. The T cell receptor will only recognise antigen if it is shown to the T cell receptor with the appropriate MHC molecule. The appropriateness depends on how the foreign antigen is processed by an antigen-presenting cell, and then presented to the T cell receptor. The processing will be described later, in Chapter 5, but the essential point is that antigens shown to a T cell receptor with MHC class I molecules require one subset of T cell to react to it, whereas MHC class II interactions use another T cell subset. These two subsets of T cells are called CD8 cytotoxic/suppressor and CD4 helper cells. The diagram in Fig. 2.2 shows the T cell receptor and its associated CD3 molecule binding to antigen, presented here in the context of MHC class II. This has occurred because the CD4 molecule on the T cell has made contact with the MHC class II molecule. The total complexity provides fail safe mechanisms. In this way T cells that have the power to start an immune response or kill infected cells will only do so if certain safeguards have been met. T cells do not normally attack the body's own cells by accident. If they do, then something has gone wrong somewhere in the making of the T cell receptor, the safeguards have failed.

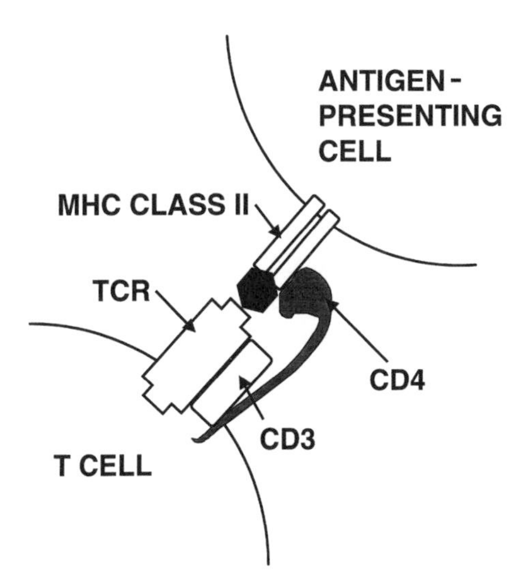

Figure 2.2. A diagram to illustrate how the T cell receptor (TCR) is closely associated with CD3, and needs to 'see' foreign antigen in the context of MHC class II antigens. MHC class I antigens would require the presence of CD8 instead of CD4 on the T cell.

When the stem cells enter the thymus, they have not yet developed a T cell receptor, and do not have the CD4 or CD8 molecules, referred to earlier, on them. The prefix CD refers to 'clusters of differentiation' antigens that are found on cells, and these have been grouped for convenience into similar types, each with a separate number. As these stem cells develop, they activate the genes that make both CD4 and CD8 and the majority of cells in the thymus have both CDs on their surface. The change from no CD4 or CD8 to both is regulated by part of the T cell receptor, its β chain. As the β chain appears on the cell surface, it forms a complex with CD3 and other molecules. This combination stops the cell from dying and rescues it for further development, as well as stopping the making of a unique T cell receptor. The type of T cell receptor that an individual cell has is now determined. In addition, developing thymocytes, that will later become the T cells of the body, must be selected for their ability to make a 'correct' interaction with antigen. Any thymocytes that might react against the body are eliminated in the thymus before they escape from it, as they would then damage other vital organs of the body.

As described earlier, T cells can only recognise antigen when it is presented to them with MHC antigens. Furthermore the MHC molecules must be the same on the reactive T cell as on the antigen-presenting cell. This restriction to one type of MHC is accomplished within the thymus during development. Exactly how this happens is still not entirely worked out, but it is known that it is achieved when thymocytes have both CD4 and CD8 on their surface, and before they mature to a form with only one type of these CD marker molecules. Again, part of the T cell receptor is important in controlling this change, but this time it is another chain, called the T cell receptor-α chain, that is involved. As the genes for the α chain are rearranged, and the newly made protein is positioned on the cell surface, it is in a position to interact with any antigen that is attached to other cells of the thymus. As part of an elaborate process of controlling these interactions in the body, and in order to safeguard the body against harmful T cell actions, these developing T cells or thymocytes are allowed to 'see' self-antigen in the controlled environment of the thymus. Then if the type of T cell receptor formed by the gene rearrangements described above has by accident produced a harmful receptor – and in this context this means one that would be able to start an immune reaction against one's own body when the T cell leaves the thymus – then the cell with this receptor is killed off. This means that all antigenic molecules on the body's cells that the T cell might meet must be 'shown' to the thymocyte's T cell receptor whilst it is in the thymus. An awesome task, and one that we still do not fully understand. However, the result is that each T cell receptor that is produced is 'tested' to make sure that this T cell will not harm the body and will only recognise antigen from invaders such as viruses. After testing the receptor and approving its form, the cell is positively selected to survive and differentiate into a T cell.

If one type of receptor made by gene rearrangement is unsuccessful, the thymocyte is allowed to try out another receptor. It seems that there is a period of about three to four days when new gene rearrangements can be tested, so in this period of time an individual thymocyte can get on with making different α chains until the cell is positively selected. This process is unique to white blood cells of the T and B subsets, and it only takes place as the cells are made in the thymus or bone marrow.

The mature T cells that are released from the thymus have developed so that they only keep either CD4 or CD8 on their cell surface and are therefore called CD4 or CD8 single positive cells. If the T cell keeps CD4 it will become a T helper cell, and thymocytes with CD8 will become T cytotoxic or suppressor cells. In contrast to this process of positive selection, cells which would be self-reactive by failing to tolerate self-antigens are deleted or negatively selected by a special process of cell elimination called apoptosis. The term apoptosis signifies a kind of cell death that does not invoke any immune reactions, unlike, for example, the killing of bacteria. The cells condense, their DNA becomes disorganised and non-functional, and the cells die. This also happens to any cells that cannot recognise MHC antigens, and they simply die by neglect – they fail to get survival signals since there are no interactions with other cells.

The majority of T cells developed in the thymus die, and only a few, perhaps only 10% correctly 'programmed' cells, can survive. When the survivors finish their development, they can leave the thymus as naive T cells. Such cells will have a T cell receptor and the CD3 molecules on their surface plus either the CD4 or the CD8 receptors. How they use these molecules to interact in immune reactions is another story to be told later in Chapter 5.

In the above descriptions, we have been detailing the development of thymocytes with T cell receptors composed of αβ chains. Almost all of the T cells in the body have the αβ T cell receptor. However, the body also contains thymus-derived T cells that have a T cell receptor made of different chains, including γδ chains. Of these, most do not have either CD4 or CD8 molecules on their surfaces, although some do have CD8, and a very few may have CD4. Most of the γδ T cell receptor T cells with CD8 are found in the gut, and almost all of the T cells of the skin are γδ T cells. Rearrangement of the δ chain gene imparts a wide degree of variation to T cell receptors amongst γδ T cells. Cells of different specificities appear to be associated with different parts of the body, and increases in their numbers may coincide with the development of certain autoimmune diseases such as rheumatoid arthritis. γδ T cell receptor T cells are also increased in number during pregnancy and may help to suppress immune responses that would be harmful to the developing baby. However, how γδ T cells recognise antigen is not so easy to understand, although some react to MHC class-I-like antigens.

The B Cell Family

The B cells are an important family of leukocytes that are made in the bone marrow before they circulate round the body in the blood. B cells are immature, and the mature form, called a plasma cell, secretes large amounts of antibodies or immunoglobulins. When the B cell develops, it makes a unique receptor of immunoglobulin on its surface. The type of immunoglobulin on the cell's surface is exactly the same as that made inside the cell, so the B cell has an advertisement on the surface saying 'this is the type of antigen I am able to bind'.

The uniqueness of the B cell receptor arises in a similar manner to that used by T cells during the development of the T cell receptor, and the processes allow a similar potential for variability. The B cell receptor is a sample of the type of immunoglobulin made by that particular cell. The process of gene rearrangement allows several million different possible B cell receptor types to be made. As with T cells, the B cell receptor of immunoglobulin is composed of 'heavy' and 'light' chains containing both variable and constant regions. Genes with the information for making the basic chains of immunoglobulin are rearranged in the nucleus of immature or pre-B cells and, at this stage, the numbers of pre-B cells can be increased by cell divisions. Once the heavy chains of immunoglobulin are made, cell divisions cease. Then more proteins are made and used to form a large type of immunoglobulin that is expressed on the cell surface. Now B cells can follow several pathways. Either they can go on having their genes for the light chain rearranged until an acceptable form is produced, or they are given a positive signal that ensures that they survive, or they may be actively deleted if they strongly interact with self-antigens. Acceptable in this context means that the B cell can only weakly recognise self-antigens and does not evoke an immune response against the body. Quite a large number, about 30%, will not have an acceptable gene rearrangement, and these die before reaching maturity. The surviving mature B cells then move away from where they were formed, out into the blood. Each B cell will have a different type of B cell receptor or immunoglobulin on its surface. Only when the B cell's receptor actually finds its antigen in the body and binds to it is there a need for more B cells with this type of receptor. The immune reaction that ensues allows these specific B cells to divide to produce many more cells with the same gene arrangements. The reproduction of cells to form identical units is called cloning, and the newly cloned cells are now renamed plasma cells. All plasma cells cloned from one original cell will make the same type of immunoglobulin, so that large amounts of immunoglobulin are produced in the body and, furthermore, this antibody is the correct type to bind to the invaders.

Human antibodies all belong to a large family of immunoglobulins (Ig) which are classified into five classes: IgM, IgG, IgA, IgD and IgE. Initially, each B cell has identical but unique immunoglobulins on the cell surface. Thus each B cell generally binds only to one type of antigen. However, at different stages in the immune response, B cells can switch from making one type of immunoglobulin to make another, or can secrete two types of immunoglobulin. All the immunoglobulins from one cell have the same variable region whatever type of immunoglobulin is produced. The uniqueness of the B cell receptor is maintained for the life of that B cell clone.

B cells also have other non-specific receptors on their surfaces which they use in the process of recognising antigen, or to obtain help for antigen recognition. These will be discussed later in Chapter 5 where the interactions between antigen and cells are considered.

Other Families of Immune Cells

Another important family of immune system cells is the macrophage lineage. This family is probably the oldest and most important of all the mobile cell families in the body. Even the most primitive animals have macrophages. These cells are called the scavengers of the body. They perform a myriad of different tasks, but a universal property is their ability to clean up the environment. This they do by engulfing and breaking down cells and their components, pathogens and inert particles such as asbestos. This activity, called phagocytosis, may have been how the immune system first started to work, but in Man these cells are immensely diverse in their actions and are at the centre of most immune reactions.

Young macrophages are termed monocytes, and in adults these are made in the bone marrow. They are small, rounded insignificant-looking cells that move out into the blood and then into tissues where they become larger cells called macrophages. Macrophages constantly patrol the environment until they meet pathogens, infected or dying cells when they are triggered into action. When this happens, they become activated to phagocytose or take in the affected cell.

Macrophages have a very large number of different receptors on their cell surface, and the numbers of these are often altered during macrophage priming and activation. None show the ability to be varied like the T and B cell receptors, and the keynote is that several of the receptors bind antigens non-specifically. Their ability to engulf particles or invaders occurs through the action of three main types of receptor that, when activated, all increase the cell's ability to take up particles and break them down. A major player in this is a receptor called the Fc receptor, which is also found on many other cells, including granulocytes which form the other main family of cells that perform phagocytosis. The binding of invaders to Fc receptors can be enhanced by a number of other factors that circulate in the blood and coat the invader to inactivate it or make it easier to destroy. Macrophages also kill infected cells.

Another important activity of macrophages is the processing of antigen. Indeed all of the 'professional' antigen-presenting cells are derived from the monocyte/macrophage line of cells. These include Langerhans' cells of the skin and the dendritic cells of lymphoid organs (see Chapters 5 and 6).

The family of granulocytes is composed of at least three main lines (Fig. 2.1): neutrophils, eosinophils and basophils, which are similar to mast cells. The simplest stem cells in the family could form any of the lines but, under the influence of factors from nearby cells, the stem cells differentiate and become more and more restricted to one line of development. Most granulocyte development in the adult takes place in the bone marrow but it can occur in many different organs, and its extent will depend very much on the needs of the individual. Granulocytes are so-called because they contain large numbers of special granules which can be released to interact with many other cells. Neutrophils are the commonest, forming up to 90% of blood granulocytes. These cells have the ability to engulf, or phagocytose, cells such as bacteria and pathogen particles, and use mechanisms similar to those described above for macrophages to take up the particles. Once in the neutrophil, some of the granules release enzymes which are substances capable of breaking up ingested particles or invaders. In particular one enzyme, called lysozyme, is very important in breaking bacterial cell walls. Eosinophils are also able to take up and destroy cells and particles but generally they have many different actions. They are particularly important for fighting protozoa (microscopic, single-celled animals) and certain parasitic worms (flat worms) that may get into the intestine. However, they can also release chemicals than can dampen down an allergic or inflammatory reaction and neutralise the actions of other granulocytes, the basophils or mast cells. Both of these latter cell types have large dense granules that contain an enormous number of active chemicals. Some of these affect how fast blood can flow through adjacent blood vessels, and others can alter the extent to which the cells in blood vessel walls are held together. If loosened, then immune system cells can all move through the walls more easily. Because the blood vessel walls become leaky, fluid seeps out from the blood, and when it collects at that site the region becomes fluid-filled or oedematous. We often see this reaction around infected cuts, insect bites and stings.

In recent years it has become apparent that, in addition to the T cells described above, there are other similar cells that lack the T cell receptor, but are not granulocytes or macrophages. One group is relatively numerous and quite important in immune reactions, although they were not appreciated or studied until comparatively recently. These are the large granular leukocytes that include natural killer cells. They have gained this name because they can kill infected cells without having a specialised T cell receptor or receptors for MHC antigens. Indeed, the lower the expression of MHC class I antigens on target cells, the better natural killer cells recognise them. Therefore, they are, to some extent, non-specific in their actions. However, they do not go around killing all and sundry, so they must have special mechanisms for identifying target cells, although, as yet, how they distinguish between different cells is only poorly understood.

The Family Retinue

Just as large family houses may be run on a hierarchical system of staff with an ultimate major domo/housekeeper/head butler who delegates areas of responsibility, so too does the body use this principle. One of the most senior organisers of both the immune and endocrine systems of the body is a line of command called the hypothalamic-pituitary axis. The hypothalamus is part of the brain. Messages from the hypothalamus reach the pituitary which is a small organ attached to the brain. This in turn sends new messages all over the body. Much of the activity of this pathway is concerned with the general metabolism of the whole body, especially how we handle stressful situations by a 'fight or flight' reaction (see Chapter 8). We now know that as well as controlling heart rate, etc. the messages from the hypothalamus and pituitary also influence the immune system. Some messages work on the endocrine glands whose hormones can alter the microenvironment of lymphoid organs. Other factors can act directly on the families of cells discussed above. Thus, although the hypothalamic-pituitary axis is not part of the immune system, it does influence some immune system activities.

This chain of influence is effective because most of the cells in each family have a characteristic set of surface receptors for factors from the hypothalamic-pituitary axis, endocrine organs or other immune system cells. Furthermore, once the receptor has been occupied by its correct binding factor, or ligand, this interaction may cause other factors, generally called cytokines, to be released from that cell. Cytokines are all short-range signalling molecules that control cell growth and differentiation, cell functioning and cell survival. There are many different cytokines and new ones are being identified all the time. However, the activities of many cytokines overlap; many cytokines require another to add to the potency of their action and few, if any, are secreted by only one cell type. So the story, like all immunology, is complex.

The cytokines have been given a wide range of names, from interleukins (always shortened to IL, and then followed by a distinguishing number) to growth factors, but they can be grouped on the basis of structure into families which, to some extent, also serve to classify their functions. The families may relate to how the cytokine structure is folded, or the shape of the receptor. Their actions are central to immune responses causing fever, metabolic changes and inflammation. One of them, tumour necrosis factor, as may be guessed from its name, kills tumour cells, activates neutrophils to become phagocytic and macrophages to kill, as well as having profound effects on many other cells. Many of the actions of cytokines will be mentioned throughout this book, and considered separately in more detail later.

Population Explosions

Anyone who has given blood during his or her life will be aware of the fact that, despite losing many cells with the donation, there is no need to worry about this in the long term. The body is remarkably efficient at realising that cell numbers in the blood are reduced, and the process of repopulation starts immediately. How constant levels of cells are obtained is not understood, but it is known that the total numbers of all blood and immune system cells are controlled. Indeed, most other organs, such as the liver and kidneys, also retain a fairly constant cell number. For example, if the liver is damaged the remaining part will try to regrow to replace the damaged cells.

Many of us have had a blood test during our life. Despite us being different from each other in size and weight, by counting and identifying the cells in the blood a haematologist can tell if there is any major blood disorder, or perhaps alert a doctor to the possibility of an infection or disease. For example, eosinophils usually make up about 2–3% of all the leukocytes in the blood, but if the patient is infected with the liver-fluke parasite (*Fasciola hepatica*) the number of eosinophils often rises dramatically. Values as high as 90% may be found. This condition is then referred to as 'eosinophilia'. This occurs because the fluke generally gets into the liver, and eosinophils are produced in great numbers to kill the flukes there. The combination of fluke damage and the increased immune reactivity all act together to cause the liver to enlarge too.

As mentioned right at the start of this chapter, the leukaemias are disorders of the blood-forming cells. Normally, in a healthy adult, there are no immature erythrocytes or leukocytes in the blood but in leukaemias this changes. Examination of a blood sample from a patient suspected of having leukaemia will often show these immature cells and so indicate the nature and severity of the disease. Since the patient has a defect in the process of blood formation in the bone marrow, immature cells escape into the blood, and this seems to upset the percentages of other cells that are normally present. Leukaemia is one of the cancers, and all cancers are characterised by uncontrolled cell replication. After most stem cells divide, they start to acquire the characteristics of mature cells. The process of differentiation depends on the presence or absence of signals from other cells or the local environment. Often the lack of a signal means that the cell cannot continue to differentiate (see Chapter 9). After cancerous cells divide, they continue to multiply and do not become specialised. Somehow the cancerous cells, as in some leukaemias, have escaped the mechanisms of cell family planning.

In the case of chronic granulocytic leukaemia, the cause appears to be a defect in the stem cells of the family of bone-marrow-derived cells. However, in other forms of leukaemia, other reasons may have brought on the condition. Radiation and some chemicals damage the packages of genes in the nucleus (chromosomes), resulting in clones of abnormal cells after cell division. It has also been realised that some cancers are started after certain viruses, such as the ribonucleic acid viruses, get into the cell and interfere with nuclear events. Fortunately, successful treatment with multiple drugs or chemotherapy can leave the least committed stem cells unaffected, and then these can repopulate the bone marrow as normal. Treatment is therefore a delicate balance between destroying the abnormal cells and preserving the normal families of cells so that they can enact normal immune responses and keep secondary infections at bay. This is why the modern approach to cancers, where the 'rogue' cells are targeted by antibodies designed firstly to identify the cancerous cells by their surface markers and then to bind to them to kill them, is so exciting. Successful elimination of 'rogue' cells would then leave the body to readjust the normal balance of cells in the blood without manipulation by doctors. The patient would of course need to give blood for routine screening from time to time to ensure that the situation had not got out of control again. Such a course of action promises to be more attractive than the long-term use of drugs, since the body is essentially working as normal again.

3

Patterns of Life

Allergies It has been estimated that, in developed societies, about one in four people are allergic to something. Most of these suffer hayfever, which is a form of allergic rhinitis, or asthma. Many allergens (i.e. antigens that cause allergies in hypersensitive people) have been isolated and described in detail. These allergens are generally antigens derived from plant or fungal proteins. However, allergy to cats is the most common allergic condition world-wide, followed by allergy to house mite dust. All babies are capable of being allergic to foods and inhaled allergens, since their immune system is just developing the ability to react in an adult manner. Generally they only receive breast milk or formula feed for several months so the reactions are not common but, if foods are not introduced slowly enough, babies up to about nine months to one year of age can show allergic responses. Then the body learns to tolerate, or not react to, the potentially allergenic substances and the baby can eat a mixed diet without reactions. About 2–5% of adults remain allergic to foods such as cow's milk, peanuts or seafoods, and about 5–10% of children may suffer asthma, with the numbers rising each year. One-third of these may grow out of asthma, but another 10% of adults are newly added to the list of sufferers. The cost of asthma to the American economy in 1990 alone was around $3.6 billion. The major stages in the response are sensitisation, activation of mast cells and prolonged immune system activity.

Allergenic substances are generally not harmful the first time the body meets them. However, the meeting has not gone unnoticed, and as the immune system responds many cells become activated until the B cells begin to produce immunoglobulins. At some stage in the process, maybe long after the allergen was met, the B cells switch to making immunoglobulin E (IgE) which, unlike the other immunoglobulins, can persist in the body for months or years. This immunoglobulin binds to the granular white blood cells called mast cells and basophils and sits there waiting for another encounter with the allergen. Within seconds of the allergen getting into the body again, it binds to these IgE/mast cell sites and activates the cell. The chemicals released from mast cell granules are potent. One substance, histamine, immediately causes the smooth muscles of airways to contract, and small blood vessels to leak. Other factors are freshly formed as mast cells release their granules, and these potentiate the histamine response. The affected region becomes red and swollen. If this effect is extensive, a potentially fatal condition of anaphylactic shock can develop. There is also a long-term effect whereby basophils and eosinophils are attracted to the area, and they in turn set off damaging immune responses at that site.

Life Begins

An active immune system is necessary from the earliest stages of life, although its role changes during our lifetime. This is a fascinating and important story, since how we cope with illnesses will be a mixture of what potential we have inherited in the genes from our parents, how the system has been programmed in early life, and what challenges it meets from day to day. We have already seen (Chapters 1 and 2) how inheriting certain major histocompatibility complex (MHC) genes can influence other events, such as the outcome of surgery, and the genetic predisposition to certain diseases later in life will be discussed more fully in Chapter 9. In this chapter we will discover how events early in life shape our immune responses, and what changes may occur as we age.

As soon as an egg is fertilised, the tiny embryo has genes from both parents. This should, in theory, present a problem to the mother. The new being is already different in composition from the mother, so the baby could potentially be regarded as foreign or non-self to the mother. Why then, is it not attacked and destroyed by the mother's immune system? We do not know all the answers, but multiple factors enter into the equation. As the sperm enters the egg and immediately after, the egg may escape destruction because it is floating free in the uterine tubes and travelling towards the uterus, where it should eventually implant. In the ovary, the eggs are surrounded by a special coat called the zona pellucida. When the egg is shed from the ovary, this coat is still present and indeed enlarged by the attachment of ovarian cells, all from the mother. Although this covering is penetrated by the sperm, nearby cells from the mother's immune system could be fooled into thinking that the fertilised egg is not foreign to the mother, and thus it would be accepted as self, and not attacked.

The zona pellucida and ball of ovarian cells are both lost when the egg implants into the uterine wall. The action of implantation is recognised by the mother and she actually starts a rejection process, which is rather like rejecting organs transplanted from one person to another. This alters the local environment around the fertilised egg, and perhaps allows it to start to embed into the uterine tissues. Without this rejection process, the embryo cannot implant into the wall of the uterus. However, once the embryo is safely implanted into the uterus, the rejection process is immediately halted, and the embryo can go on to develop a network of blood vessels that intertwines with the mother's but is entirely separate. In this way, the oxygen can reach the baby's tissues and waste products can move across into the mother for elimination. The placenta is where the two blood systems are closest, and the cells lining the blood vessels actively screen all substances passing between the mother and the baby. In addition to oxygen being passed across the placenta, beneficial molecules, such as nutrients, are also let across. The placental cells also block the entry of any maternal immune cells, such as T cytotoxic or killer cells (see Chapter 2), that could cause harm to the embryo. The outermost cell layer of the placenta is called the trophoblast and this, unlike most cells of the body, does not have any classic MHC class I or class II antigens on the cells. This is unusual, but the brain is another exception. As with our brain, this reduces the potential for the immune system to recognise and act against the unmarked cells. Thus in pregnancy, the maternal immune system cells cannot harm the placenta, since without MHC class I molecules the cells cannot be recognised as foreign as described in Chapter 1. Even if any antigens from the embryo escape out to the trophoblast cells, they cannot be presented to T cells of the body without the MHC class I markers being present, so no immune response is evoked.

Both the mother and the foetus towards the end of pregnancy produce antibodies. Those of the embryo could possibly get beyond the trophoblast into the mother, but they do not appear to cause harm. Even if the mother's cells get into the embryo, which only rarely happens, there should be no problems, as the baby's immune system is not functional until very late in pregnancy. At the time of birth, however, there is a problem if the erythrocytes (red blood cells) of the mother and baby have different Rhesus antigens. The term Rhesus, which is generally abbreviated to Rh, comes from the rhesus monkey which was used in the identification of this set of antigens. In Western Europe, 85% people have Rh antigens on their erythrocytes, so they are described as being Rh-positive. The rest are Rh-negative since they have no Rh antigens. The percentages are different for people with African or Oriental origins. The people who do not have these antigens on their erythrocytes have a 50:50 chance of reacting against the blood of Rh-positive blood donors. But in pregnancy there are serious risks to the baby if the baby inherits Rh-positive antigens from the father and the mother is Rh-negative, as the baby could develop haemolytic disease of the newborn. The first time a Rh-negative mother becomes pregnant with a Rh-positive baby there are normally no problems. But at birth, and sometimes towards the end of pregnancy, some of the mother's and baby's blood can mix if the placenta tears, and the mother will then make antibodies against the Rh-positive antigens from the baby. Now the second time the Rh-negative mother becomes pregnant with a Rh-positive baby, the mother's body already contains B cells that can be stimulated to make antibodies that would destroy the baby's Rh-positive erythrocytes. Antibodies can cross the placenta in late pregnancy, and would mix with the baby's blood if the placenta is damaged then or during the birth. If this occurs the infant can become very sick with hemolytic disease. Then it is sometimes necessary to completely transfuse the Rh-positive baby with new Rh-positive blood which has not been contaminated with antibodies raised to Rh-positive antigens. It is possible to overcome this problem when there is a mismatch of Rh-factors by injecting the mother with immunoglobulin, known as anti-D, that is formed against the Rh-factor within three days of the delivery of a first Rh-positive baby. This globulin mops up the Rh-positive antigens from the baby, and they are disposed of before they can alert the mother's immune system into a full reaction. This also has the effect of protecting the next Rh-positive baby.

The New World

At birth, the entire region around the baby changes from a privileged and highly protective environment, to the rigours of the outside world. The baby is still growing fast and not all of its systems are fully functional. In particular the gut is not able to take in and use the wide variety of foods that adults eat. This is fortunate, because the baby, by feeding primarily on breast milk, gets a lot of help to survive the new environment. The first milk, colostrum, is scant and yellowish, so many mothers do not attach any value to it. However, it has important functions in the early days after birth. It contains many cells of the mother's immune system, especially eosinophils and macrophages, that can engulf and destroy bacteria. The macrophages also produce many soluble factors including lysozyme, interferon and fibronectin. Lysozyme kills bacteria directly by breaking open their cell walls. Interferon is a powerful anti-viral agent, and fibronectin encourages macrophages to be aggressive and to kill any pathogens that get into the baby. All of these factors are necessary as the baby's gut is not fully developed and any bacteria that get into it are really damaging. Indeed, in developing countries with a high infant-mortality rate, bottle-fed infants are at least 14 times more likely to die from diarrhoea compared with breast-fed babies. Even in westernised societies, premature babies on formula feeds are at a greater risk of dying from inflammation of the intestines than their breast-fed counterparts.

If the baby is allowed to suckle, breast milk begins to flow after about one to two days. Mother's milk is invaluable to the baby. It supplies nutrients which the baby's gut cannot manage to digest from solid food, as well as a wide range of molecules that give the baby immune protection, favour the development of strong defences against viruses and bacteria, and speed up the full development of the gut and respiratory systems. The milk continues to supply the baby with many factors previously obtained across the placenta. Such factors, including maternal antibodies, are necessary to protect the baby, as the immune system of the young child is not fully formed until about five years of age. Infections of the gut, chest, ears and urinary tract are better controlled in breast-fed babies.

Breast feeding for at least six months is recommended, and continuing for up to two years or even longer, gives the baby a better chance of being a healthy adult. Modern life styles make these longer breast feeding times difficult to achieve, and nowadays mothers are often encouraged to stop breast feeding after only a few months. Nonetheless, it has been shown that breast feeding a baby for at least six months protects against eczema, as well as food and respiratory allergies in young children. One recent study has shown that children of 17 years who had been breast-fed for longer than six months still had better responses to allergens than those breast-fed for shorter periods of time. It is suspected too that there are more dental caries and dental problems in those who were bottle-fed as babies, and probably the greater incidence these days of cow's milk allergies and intolerance to its products could be the result of too early an exposure to these substances. Breast-fed babies also obtain more of a certain type of nutrient essential for the development of the nervous system, called long-chain polyunsaturated fatty acids, which are generally absent from formula feeds. It has been suggested that since the brain needs these fatty acids to develop well, this could be the reason for those who were breast-fed as babies having a slightly but significantly better intellectual development. The mother also may benefit from lactating, since it does tend to reduce fertility and may prevent ovarian and premenopause cancers.

A very important component of milk is an antibody or immunoglobulin called secretory IgA. Pairs of these immunoglobulins can be taken into cells that secrete milk by binding to a receptor on the cell's surface. Once inside the cell, the receptor keeps the two IgAs as a unit, so that when the complexed IgA is secreted into the milk the IgA structure is stabilised. In this way, the IgA cannot be destroyed, and it acts as an effective protection against infections entering the baby. In adults, IgA is secreted from cells that line the gut and respiratory system. Bacteria get into the gut with food, and everyone has bacteria in their gut. Even the newborn baby's gut rapidly acquires bacteria. Some are useful, as they break down ingested food, whilst others are harmful and cause tummy upsets and diarrhoea. In the small intestine, in some as yet unknown manner, IgA can distinguish good from harmful bacteria and it only breaks down the unwanted ones. In addition, some factors from the mother's milk actually help beneficial bacteria to grow. For example, a factor called bifidus encourages the *Lactobacillus bifidus* bacteria, which we all need in our gut. IgA's control of bacteria in the gut does not cause inflammation. This is important as it would damage the delicate and underdeveloped gut walls of the baby, and cause irritation in children and adults alike.

First Challenges

Most people agree that babies should only be given breast milk or formula feed at first, and other foods are introduced much later and slowly. Although the immune system is not fully formed at birth, the baby's T and B cells in the body can already make some weak reaction against challenges, and indeed it has been suggested that food allergies could even start before birth. Certainly at the time of birth, people have demonstrated that the baby's blood cells can be sensitive to egg proteins and inhaled allergens such as pollen and house mite dust. Children born in the pollen season are much more likely to develop hayfever later in life, and children born in the autumn similarly are more affected by house mite dust and the skin and hairs of animals.

By about three months of age, all babies have the ability to react to food and inhaled allergens. These responses are strongest around nine months of age and then wane very rapidly. It is as though all new foods and air-borne challenges make the body react during these early months, but the body, over the first year, learns to recognise them and becomes accustomed to them, and eventually stops reacting to them. This ability to accept and not react to challenges is called tolerance. Tolerance is a very important concept for the immune system. It is the healthy (or correct) basis for not reacting to our own body's cells, but it is also the reason for some diseases gaining hold on us. In some cases, the body can incorrectly tolerate certain invaders or cells within the body, which are then free to do damage. The persistence of cancerous cells has been classified as a form of tolerance. If only the body would recognise them as harmful, then they could be acted against by the immune system. With regard to food allergens, tolerance is necessary. As the baby is slowly introduced to new foods, so its body becomes tolerant to them, and by about one year most babies will accept and not adversely react to a wide variety of foods. The process happens slowly, and excess of any one product can cause problems. Thus some babies may get allergies to proteins in the white of chicken's eggs, and cow's milk allergies may be related to too early or too great an exposure as a baby. Fortunately as tolerance develops, most babies lose this allergic response and tolerate all foods. Nevertheless, some people inherit a tendency to become more sensitive to allergens, and once allergic to one food they can rapidly become sensitive to others, often sea-foods or nuts (see Chapter 4). This susceptibility to be affected by allergens in food or from the air is called atopy. Estimates of how many people are born atopic are generally around 2–5% in westernised societies although, because atopic people are also more likely to react to environmental factors and this type of allergy is on the increase, some researchers think that the percentage could be as high as around 20% of people. Since these reactions could be exacerbated by too early an exposure of babies to allergens, once again the importance of breast feeding for at least four to six months and preferably longer, even up to nine months, is obvious for the healthy future of the child.

Whilst the young baby is within the home environment, it does not encounter a very wide range of pathogens; however, if brothers and sisters go to school, or when the child itself starts school, a whole host of immune challenges present themselves. As explained before, some of the mother's immunity can be passed onto the baby through breast milk, and this lasts for several months. These are antibodies made by the mother, but the child needs to make its own, and to do that it must meet antigen derived from the pathogen. Since 1775, when Edward Jenner established that milkmaids who were exposed to cowpox did not get smallpox, the principle of deliberate immunisation has been developed throughout the world. In the case of smallpox, the closely related cowpox vaccinia virus was injected in 1776 into an uninfected boy who reacted by the production of cells making antibody to the cowpox virus. These cell types were memory B cells that remained in the body, so that the next year, when the boy was inoculated with smallpox, the body's memory cells for cowpox were fooled into thinking that the smallpox virus was cowpox virus and antibodies were quickly made to wipe out the infection before it got going strongly. The boy did not develop the disease. Today the disease of smallpox has been eliminated from the world by vaccination programmes.

The value and implementation of vaccination programmes throughout the world and the selection of effective immunogens will be considered later in Chapter 10. Here it is important to realise that the young baby can benefit from such vaccinations because the baby has not yet met any of the expected pathogens. In most developed countries vaccinations against polio, diphtheria, pertussis, tetanus, tuberculosis and rubella (or German measles) are commonly offered, and in addition vaccinations against mumps, measles, hepatitis B and bacterial influenza are commonly given.

Vaccinations can be given from about one month of age onwards, when the value of the inherited mother's immunity begins to wane in bottle-fed babies. Breast feeding can result in more antibodies being produced in response to an immunisation, although in the very young vaccinations can be ineffective. Thus nowadays, in Europe and America, vaccinations against diphtheria, tetanus, pertussis and polio are given first between two and four months of age, and the others, as well as booster vaccinations, are given later, in some cases up to the age of 18 years. These regimes are varied according to the needs of the population. In Asia, for example, where hepatitis B is endemic in the population, babies are immunised from birth.

Adjusting to Life

Childhood is a period of intense learning. The immune system is learning too. When a child starts school, it has good protection against some of the commonest diseases due to meeting them as a toddler, or being vaccinated. Despite this, there are still many new bacteria and viruses that the child will meet, and during this period its immune system is developed by these encounters. Typical causes of pain and illness in children are infections of the tonsils, adenoids and appendix, which swell and become tender with immune challenge. The incidence of problems from these organs peaks in the early teenager, and declines in adults. An enormous number of T and B cells is produced from the thymus and bone marrow in childhood, and both organs are at their fullest development shortly after birth. The new T and B cell repertoires that these organs make circulate through the body, ready to fight invasions. Slowly in childhood, and to some extent throughout the rest of life, as challenges are met, the body becomes tolerant to the new world, so the cell-producing parts of these organs get smaller and replaced by fat. They always remain capable of producing some new cells, but when that ability is severely compromised then diseases may gain a hold, as with secondary infections after the human immunodeficiency virus (HIV), which causes the acquired immune deficiency syndrome or AIDS, infects the body.

Allergic reactions are not diseases, and they can occur in many otherwise healthy people. They are called immediate or type I hypersensitivity reactions. Types II–IV reactions will be discussed later, as appropriate, but the haemolytic disease of the newborn, already discussed, is an example of type II hypersensitivity. Many type I reactions are against food allergens, and hayfever and asthma are also included here. In allergic reactions, the body becomes sensitive to the tiny particle of allergens through the over-production of the IgE type of immunoglobulin. The way the body reacts to peanut allergens illustrates how this happens, but the intense and serious reaction, called anaphylactic shock, which can occur in people already sensitised to peanuts is not common with most food allergens.

Peanuts are one of the commonest causes of severe allergies both in children and adults in the USA and France. In the UK in the early 1990s, it was estimated that there were 20,000 sensitised children. Strangely, for some unknown reason, peanut allergy is very rare in some countries such as Thailand, Egypt and India. However, this may be because in the West some babies were unintentionally exposed to peanut oil. Some creams for dry or cracked skin contain arachis oil from peanuts. When breast feeding a baby, the nipples can become cracked and sore and some mothers use creams to ease this painful condition. Exposing the baby to peanut oil so early in life could sensitise the baby to peanuts later in life. Even formula-fed babies were discovered to be at risk from peanut allergens, this time given in another form. Until recently, peanut oil was used in many baby feed formulas, especially in France. Another risk, usually to older children, is that peanut oil is sometimes used as a carrier for antibiotics such as penicillin when they have to be injected into the body. Most people who react to peanuts have already shown less serious reactions to other foods, especially milk and eggs. So now health-care workers advise mothers to avoid any potentially allergenic substances and only to introduce babies to new foods slowly and in small amounts. If one kind of food is given in excess, it appears easier for allergic reactions to start. Once started, they are hard to eliminate.

The first few times peanuts are contacted, the person is not aware of any reaction, although inside the body the immune system of the skin or lymph nodes acts normally to cause B cells to produce antibodies of the IgD and IgM type, in order to combat this strange, non-human protein. In normal reactions, the release of immunoglobulins is often assisted by certain T cells that have already bound the antigen. We call these cells sensitised or acti-

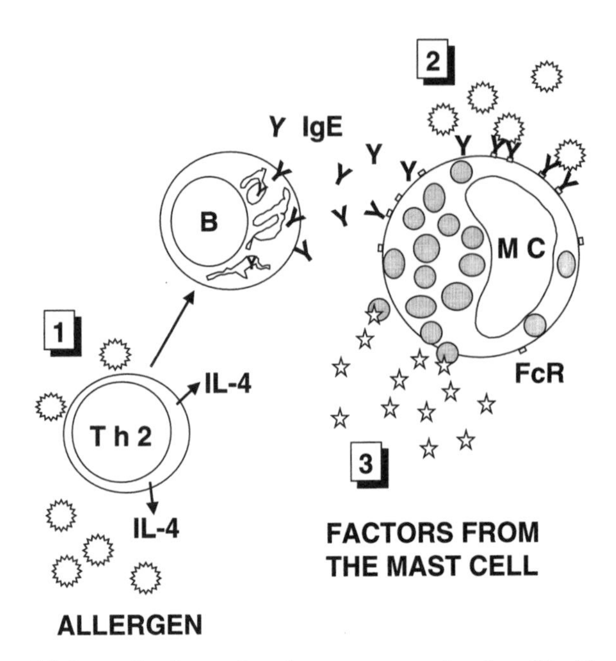

Figure 3.1. In an allergic reaction, the new contact of antigen [1] with a sensitised cell that had previously met the antigen (here the Th 2 or T helper 2 cell) causes it to release interleukin-4 (IL-4) and other cytokines. These facilitate the release of immunoglobulin E (IgE) from B cells. This IgE binds to Fc receptors (FcR, small boxes) on mast cells (MC) [2]. When antigen meets the mast cell it cross-links two Fc receptors and this makes the mast cells release histamine and other factors [3] that dilate blood vessels and start the allergic reaction.

vated T cells. They give their help in the form of locally acting factors, or cytokines, that can speed up or direct the reaction of the immunoglobulin-producing B cells. These T helper cells are shown in Fig. 3.1, where contact with peanut protein has caused them to release interleukin-4. This powerful cytokine makes the B cells change the type of immunoglobulin it releases to IgE (shown as a Y in the diagram). IgE is rapidly mopped up from the blood stream by mast cells which can bind it very easily. In a non-allergenic person this second reaction whereby IgE is produced is stopped fairly quickly by T suppressor cells, and that is the end of the story. However, in some people – perhaps because of their genetic make-up, or because they have a viral infection at the time, or because there is a very large amount of allergen – the reaction is not stopped. The body is left after contact with peanuts with many mast cells that have IgE on their surface. Since the IgE was produced only in response to peanut antigen, these mast cells are sensitised to peanuts. So unfortunately, when the person now eats peanuts or even smells them and peanut antigen gets into the body, as soon as peanut antigen comes near sensitised mast cells, that is mast cells with 'peanut antigen' IgE on their surfaces, the new antigen rapidly brings two IgEs together and binds to them (stage 2 of Fig. 3.1). The binding stimulates mast cells to release their potent contents (stage 3 of Fig. 3.1). This is why suddenly one day, when eating peanuts (or even just entering a room that contains peanuts), certain sensitised people have a violent reaction that involves their whole body, causing great distress. The speed of the reaction is due to the fact that the allergens are passed around the whole body in a matter of minutes, as the heart is capable of pumping the entire circulating volume of blood (approximately five litres) around the body in one minute, so in this time all the mast cells with IgE on them in the whole body are exposed to the peanut allergen. The problem is that there are many mast cells, all in the connective tissues around blood vessels and major airways, and they are all activated at once. The chemicals released cause smooth muscles in blood vessels and airways to contract quickly. The peanut-sensitive person suddenly feels short of breath as the airways contract, and flushed as the whole pattern of blood flow changes to every organ. This can happen so violently and fast that a person can become comatose and may die unless treated immediately. This is the very serious condition called anaphylactic shock. People who are known to be sensitive to peanuts should always have a syringe of adrenaline available to counteract the reaction. It is also possible in some 'at risk situations' to minimise the chance of exposure to the allergen. For example, if a person with an allergy to peanuts is going to fly, they should alert the airline carriers some days before travelling, to ensure that no peanuts are served on the flight, and that the cabin air is sufficiently renewed before travel begins. Indeed, the seriousness of this reaction has been appreciated by some airlines, so that many airlines no longer give out peanuts for this reason. Fortunately these conditions of the whole body are very rare, although similar but far less severe reactions can occur with other allergens. In these cases, the reactions are often over parts of the skin and the reactions often die down fairly quickly. These do not endanger life.

Hayfever is a similar involvement of IgE, but the reaction seems to be localised to the nose and mouth where the pollen gets in. Especially at the start of the pollen season, the reactions can involve the local lymph nodes, so hayfever sufferers may have swollen glands in the neck and run a fever. Some people grow out of hayfever as they get older, but there are some adults who develop it later in life. Generally anti-histamine reduces the effects of IgE on mast cells, but regimes of exposing the affected person early in the year to ever-increasing doses of the specific allergen to induce tolerance is often tried in order to reduce the sensitivity to pollen or any other specified allergen. The type of allergen, such as grass or birch pollen, must be clearly identified for this approach to work.

As the body become mature, so other changes occur. With the onset of puberty in women, menstrual cycles begin. These are created by cyclical changes in hormones, and most of these hormones will alter immune responsiveness. It is a highly complex pattern of change and the effects of the hormone cycles have not really been explored, except in relation to women with breast cancer. For example, it is now known that anti-breast-cancer treatments can be more effective at some stages of the menstrual cycle than at others. Although we think of the breast tissue in women as being greatly changed only during pregnancy and whilst breast feeding, some women are aware of the fact that monthly changes in breast tissue occur quite normally in all women whether pregnant or not.

Teenagers suffer from altered immune responses as a result of changes in sex hormones around the time of puberty. The indignities of acne are a severe embarrassment to many young people. With the increase in sex steroids in the pubertal teenager, the fat glands of the hairs, known as sebaceous glands, become over-stimulated and the region can become infected if the ducts become blocked. The blocking agent is a mass of cellular tissue from the skin itself, and so it contains melanin which gives the acne a black appearance, called a 'blackhead'. The fatty glandular secretions continue to be produced, the region swells as the outflow is blocked and immune system cells move in to clear up the mess. The region can become infected with a bacterium that breaks the fat down. Unfortunately, the breakdown products are extremely irritant if they escape with the fluid exuded, or if the blackhead is ruptured. Since around puberty young men are beginning to grow beards, shaving exacerbates the problems. Fortunately, the problem usually does go away by the mid-twenties, and there are very effective tetracycline drug therapies that are not only able to control the acne, but can also reduce the resultant scarring to a minimum.

With puberty comes the possibility of a woman becoming pregnant, and this causes quite a few changes in her immune responses. The mother's immune system changes during pregnancy, perhaps to stop the foetus being rejected since it is a new person and as such is non-self. In general, all of the immune reactions that use T cells primarily to kill foreign or infected cells by cell-mediated immunity reactions are weaker during pregnancy, and the humoral immune reactions using the B cell system are stronger. The B cells, being the lineage that produces antibodies to pathogens, are of benefit to the foetus, as many of the antibodies in the mother move across the placenta to protect the unborn baby. This is especially useful as the mother's antibodies have been made against the pathogens in her environment. When the baby is born, the baby already has some protection in its system to identify and react against its new environment. The protective effect of maternal antibodies in the embryo can last for weeks or even months after birth. Normally any antibodies that would harm the baby do not cross the placenta.

Because cell-mediated immunity in the mother is weakened during pregnancy, it can allow other harmful events to intercede. It is well documented that many infections, such as those where bacteria need to be killed by T cells, or conditions such as malaria, leprosy, toxoplasmosis or the infections associated with HIV, can flare up in pregnancy. Some autoimmune diseases, for example systemic lupus erythematosus, multiple sclerosis and several others, all get worse. It is thought that the enhanced humoral immunity of pregnancy results in increased numbers of harmful autoantibodies circulating in the mother. Curiously other autoimmunities, particularly rheumatoid arthritis, get better during pregnancy, only to recur afterwards.

Some mothers who had normal, healthy lives before becoming pregnant become diabetic during pregnancy. This form of diabetes is called gestational diabetes. Like other forms of diabetes mellitus there is a deficiency of insulin which reduces the use of carbohydrates in favour of fat and proteins and alters the composition of the urine excreted so that more sugars and electrolytes are lost than normal. These changes affect the unborn baby. During pregnancy, the baby is often heavier than expected for its age. At birth there may be complications with breathing and the excretion of urine because the respiratory and kidney systems are damaged. Congenital abnormalities also occur, and the severity and extent is closely related to the severity of diabetes in the mother. About 10% of babies born to mildly diabetic mothers may die, and death rates can be as high as 30% in severe gestational diabetes. The diabetes usually goes away after pregnancy, but may recur later in life as late-onset diabetes. With good ante-natal care such problems can easily be identified and successful measures undertaken to control the diabetes and ensure a healthy baby.

How the changes in maternal immunity during pregnancy are brought about is not yet understood. The mother's body, especially around the developing embryo, contains more T suppressor cells and natural killer cells. These may help to keep in check any immune reaction against the foetus. At the same time the body allows the development of more T helper cells of a type that direct the immune responses towards humoral immune responses. These are called T helper 2 cells, and they produce cytokines that also help the fertilised egg to implant, and the pregnancy to be maintained. Indeed it has been found that if T helper 2 cells are low in number in a pregnant woman, then spontaneous abortions may occur. Levels of all the steroid hormones such as oestrogen and progesterone are greatly increased in the mother during pregnancy, and there is also an increase in other factors from the ovary, adrenal gland and the placenta, all of which are powerful controllers of immune reactions. Fortunately increased blood levels of one of the steroid hormones, corticosteroid, allows more T helper 2 cells to be produced. These steroids and other factors, in non-pregnant adults, control the immune responses of the whole body by acting through the brain and central nervous system to alter the levels of many different hormones in the blood. In addition, during pregnancy the placental and uterine factors act on the placenta to beneficially affect the development of the baby in the uterus.

During pregnancy it is also probable that changes within the T and B cell-forming sites of the bone marrow and thymus play a part in modulating the immune responses. Whilst the bone marrow changes have only recently been described, farmers, veterinary surgeons and research scientists knew many years ago that pregnant animals have a small thymus containing only a few developing thymocytes. This state continues after the baby is born, but immediately after lactation ceases the thymus enlarges and becomes full of developing thymocytes again. In animals though, it has been found that the thymus does not totally reduce its activity as the central part becomes larger and accumulates T cells there. Thus it seems that fewer thymocytes are produced during pregnancy, and those that are may have a different function. This might help to explain the appearance and disappearance of pregnancy-associated diseases.

In animals there are also marked changes in the immune system with breeding, which often occurs seasonally. Whether it is breeding alone or influences from the changes in day length and temperature that bring about these alterations in immunity is not known. Changes in temperature, daylight, available food, animal activity and endogenous hormone rhythms have all been shown to influence immunity in experimental conditions, but they may not have the same effects in the wild. Even Man, who may think that he/she is relatively unaffected by many of these factors, has a seasonal variation in the numbers of circulating leukocytes that is imposed on the normal daily rhythm.

The daily fluctuations are controlled through the eyes, a special region of the brain and a small organ in the brain called the pineal. There seems to be a natural rhythm in the brain which is reset according to day/night cycles and is enacted through the substance melatonin, or so-called messenger of darkness, from the pineal gland. Melatonin, the hormone controlling, amongst other things, the black/brown melanin pigments in the body, is synthesised and released at night. It is essential for synchronising hormone activities with environmental signals. Its importance to disease in Man is just emerging. It appears to be able to inhibit the development and growth of cancers and to correct immunodeficiency states resulting from acute stress, viral diseases, aging and drug treatments that damage the early cells of the immune system especially those used in cancer therapy. Since sleep is generally induced by infections, perhaps one should give in more often and let sleep factors work to combat diseases.

An awareness of the influence of day length has been highlighted by the recognition of the condition called SAD (seasonally affective disorder) and the problems of coping with time-zone changes in air travel. Patients with SAD are helped by exposing themselves to bright lights during the winter months in the Northern hemisphere, when the day length is so short. Signals through the retina reset the normal cycles, and the patients suffer fewer depressions and have better health. In the case of travellers, it has been suggested that formulations of melatonin can be used by air travellers to maintain their normal rhythms of day and night, and so cope with short-term disruptions. However, the effect of increased melatonin levels during daylight hours has not been fully tested yet, and such treatments must be fully evaluated before being generally available.

The Later Stages of Life

With increasing age, there are fewer pathogens that are new to an individual, and the immune system becomes less active in producing lots of new cells. Furthermore, after the menopause in women, and similarly timed but less noticeable changes in men, the levels of sex steroids change and this influences the immune response of the whole body. As immune responsiveness decreases so there is an increase in the incidence of infections, autoimmune conditions and cancer. All of the immune system changes are subtle and interactive, and no single overriding important factor has yet been identified, although recently genes associated with ageing have been recognised.

Compared with the young, older people have stem cells of the T, B and macrophage lineages that make fewer effective cell divisions. If damage occurs to DNA as cells divide, the faults cannot be so easily repaired in older people. Thus the demand for healthy new immune system cells in disease cannot be met so readily as in the young. Older, chronically ill patients may not be able to produce enough leukocytes to cope with disease. Also, even if the cell numbers are normal, sometimes the cells themselves do not respond so well; for example, inflammation in old age cannot be cleared up quickly since neutrophils may not be as good at engulfing debris and infected cells as they are in young people. Although macrophages may continue to engulf debris, their other functions of processing antigen for presentation to T and B cells and the secretion of cytokines needed for immune interactions are reduced. B cells are just as numerous, but the subpopulations are different and their ability to react with T cells is often reduced in the elderly. Although memory B cells can react well to antigens, new challenges are more difficult to combat. Indeed it is more difficult to immunise older people against important killers of the elderly such as influenza. Diseases such as tuberculosis can appear in people over 70 years of age. The numbers of T cells do not decrease very much with age, although their capabilities may decline and there is an important shift in the types of cells present so that there are fewer CD8 T cells than in younger people, and the CD4 helper cells are not so effective. From the few studies of 'survivors', it appears that they have a degree of immune responsiveness normally associated with younger people. A healthy mind, body and immune system all help people to live for a long time.

Travelling the World

One of the exciting aspects of modern life is the greater ease of travel, not only around one's own country but to completely new environments. With the new environment come new immune challenges, and it is common practice to advise travellers to be immunised against the commonest and more troublesome diseases they might meet. The problems of almost 50% of travellers going from a temperate to a tropical climate is diarrhoea, caused in most cases by a bacterium called *Escherichia coli*. This is present all over the world, including temperate regions, but if the water supply is polluted by animal faeces, high numbers of the bacteria can be present and then it can cause diarrhoea when the water is drunk without it being boiled or treated chemically. Many other organisms also cause diarrhoea in infected people including *Giardia*, *Entamoeba histolytica*, various viruses, especially rotaviruses and adenoviruses, and bacteria such as *Shigella*, *Salmonella* and *Campylobacter*. These are all generally ingested in polluted water, maybe on unwashed salads, or as a result of lower standards of hygiene in handling and processing food. Thus the traveller can, to some extent, guard against contracting these diseases by carefully selecting what is eaten.

On arriving home, the commonest cause of fever in travellers is malaria. This is one of a completely different range of diseases, caught from parasites. Other troublesome diseases include yellow fever, liver fluke, scrub typhus and river blindness. The diseases are generally carried by another animal, often mosquitoes or flies, and are passed to Man when the insects bite. It is important to try to avoid being bitten by insects and several simple measures can be quite effective. Mosquitoes bite at night, so sleeping under protective bed-nets, and wearing long-sleeved shirts and trousers in the evening help. It is also reputed that eating garlic or garlic capsules that are odourless, several days before entering a malaria area and during the stay there, imparts protection from being bitten. Probably the garlic waste constituents are passed onto the surface of the skin in the sweat, and this may act as a deterrent. Certainly strong perfumes, especially those associated with flowers, can increase the likelihood of being bitten. Many travellers though take drugs beforehand to reduce the effect of the parasite. This is termed a prophylactic use of drugs. In the case of malaria however, the parasite is able to adapt and become resistant to the prophylaxis used. Therefore the traveller must seek medical advice on the best prophylactic treatment before entering each new malaria-infested region.

Children under five years of age are greatly affected by malaria and its more serious form of cerebral malaria that infects the brain. Often children do not have an adequate immune response to tackle the infection, and some do not develop a high fever so parents are not alerted to the seriousness of the infection. However, children born into areas where parasites abound can become immune to malaria even without showing any signs of infection. Older children and adults can raise an immune response, but this does not occur where the level of infection is low or seasonal. Pregnant women, even if they do already have good immunity, are more liable to have a severe infection. This causes the weight of the baby to be low at birth, and the baby may die soon after birth. This is probably due to the fact that the erythrocytes in the placenta are greatly infected, and are not as efficient at carrying oxygen, so the structure of the placenta is altered. This limits the exchange of nutrients to the developing baby so it does not grow as well as it should.

Newly arriving visitors, primed with their prophylaxis tablets, realise that the local people do not take such drugs, and may decide that they will not need protection either. This is not so, since the visitor's immune cells have never met the malaria parasite before and infection can be serious. Similarly, long-term workers in a malaria zone should continue to protect themselves as long as they are likely to be bitten by mosquitoes. Many people do not wish to continue the prophylaxis drugs all the time though and, since the greatest risk of being bitten is in the rainy seasons, some people discontinue the drugs in the dry season. However, then extra vigilance is needed as medical treatment should be immediate if the person contracts malaria.

Another result of rapid intercontinental travel is that diseases can be imported into countries that are normally free of the conditions. The local people are then very susceptible since they have not had the chance to raise any antibodies to the disease. Historical accounts document how the arrival of western Man into Africa, Central and South America and Australia brought in new diseases against which the natives had no protection. Some of these had disastrous effects on the population and caused widespread illness and deaths in people who were, in other respects, well able to cope with their own diseases. Also well documented were the illnesses, often fatal, that travellers and explorers contracted in their new environments. Modern air travel makes these situations arise more quickly. The very rapid spread, in perhaps just a few days, of new forms of influenza virus has been observed over distances as far apart as Australia and Europe. Space travellers are aware of the possibility of taking diseases and viruses out of the earth's environment and of polluting previously uninfected extraterrestrial lands. Conversely too, space travellers are at risk of picking up pathogens, should life exist on other planets, and of bringing them back to earth.

Of current concern is the spread by travellers of the HIV virus that causes AIDS. Men working away from home may be at greater risk of contracting HIV, and can of course carry the virus back to their home town. The spread of HIV in America started in the 1980s in the port towns of New York, San Francisco and Los Angeles where it may have come from Africa or some other destination such as Haiti. Within a few years it had been taken many miles by infected persons. Gay men and intravenous drug abusers were particularly affected. Some of these in their travels spread the virus rapidly by frequenting places in the new towns they visited where intravenous drug abusers or homosexuals gathered. It was inevitable that pockets of infected persons sprang up along highways and in large conurbations. In some cases the pattern of disease-spread could be followed by tracing the movements of one person, or small groups of people. Such new diseases of mankind are much harder to control than the older bacterial diseases which Man has, over the years, devised the means of surviving. Sadly in the case of HIV, the body cannot easily overcome the virus, and therefore it is difficult to cope with the myriad of other infections that arise when HIV has reduced the immune system to a very low level of competence. When immunocompetence is severely compromised in this way, other infections that a healthy person can readily fight and overcome cannot be resisted, and AIDS is typified by a range of such diseases taking hold.

The human body, over the years of its existence, has learned to survive many immune challenges presented by nature and by civilisation. The baby, although vulnerable to disease and infection, is well protected by the mother, and the slow introduction to new environments allows the time for immunity to all common diseases of the local environment to build up. However, modern life has now imposed its own problems. Not only does Man travel more, and risk exposure to new diseases to which immune defences have not been developed, but additionally the survival age in most populations is rising fast. Older people are therefore more vulnerable in the population to these new diseases, whether brought into their own country by travellers or encountered by ageing globe-trotters. Fortunately, communications improve every day, and the movement of diseases round the world can be monitored so that drugs can be prescribed to help meet the challenges.

PART II – REACTION

4

Repelling Invaders

Psoriasis This skin condition, which affects about 2% of the population, often develops in people between the ages of 20 and 30 years who have a certain genetic disposition to develop the disease. This is because of their major histocompatibility complex (MHC) genes in the Cw6 region of chromosome 6 – see Chapter 1. Large scaly lesions form, particularly on the skin of the elbows, knuckles and knees. In severe cases much more of the body becomes affected, so the patient finds it cosmetically embarrassing. About 10% of patients suffer a form of arthritis which flares up as the psoriasis is controlled. At each affected site, the skin cells proliferate rather faster than elsewhere and develop in a slightly different manner, and there are changes underneath that include alterations to the fine blood vessels under the skin, allowing the lesions to bleed easily. Thus the normal skin barrier to the body is weakened at these sites, and bacteria often live in the lesions. One bacterium called *Staphylococcus aureus* is quite harmful if it gets inside the body. Surgeons are careful to avoid cutting into such psoriatic lesions to prevent the spread of the associated bacteria. The treatment of the disease is difficult but ultraviolet light lessens the severity of the lesions. In some cases, the anti-cancer drugs can be used since they generally reduce cell proliferation and the lesions shrink.

Major Defences

The first and most important protection against invaders is a good barrier to their entry. This is afforded by the skin which is tough and comparatively impenetrable. In addition, the layers of the skin contain cells of the immune system that are specialised to handle invaders, and all of the skin can be patrolled by mobile immune system cells. Nevertheless, the body must be able to take in air, food and fluids, and let out waste substances, so the barrier is incomplete. This chapter will describe the ways in which the body is designed to hinder pathogens from attacking the body. Often there is a series of barriers that comes into play, depending on how far the pathogen has penetrated into the body. However, even the simplest of defences are quite intricate.

The skin is the largest area of the body exposed to invaders. It is composed of two main parts: a tough waterproof, flexible outer region and a supporting layer underneath. The outside covering is a multilayered series of cells collectively called the epidermis. This is the surface that we see when we look at the skin. Where the epidermis is very thin, as over the backs of the hands, it is so thin that blood vessels can easily be seen through the semi-transparent epidermis. There is a layer of connective tissue with blood vessels and nerves underneath the epidermis, and below that numerous fat pads. Sweat glands and hair follicles are mainly in the connective tissue beneath the skin, but they both penetrate the epidermis to reach the surface. The epidermis itself has no blood vessels or drainage channels for white blood cells and fluid (lymphatics) so all its nutrients must diffuse across from underneath. Thus the epidermal cells, by being tightly joined to each other, form a continuous multilayered sheet of cells that completely surrounds and protects the body.

The cells of the epidermis are constantly being renewed. The dividing cells are deep in the skin, closest to the connective tissue and are therefore protected a little from the rays of the sun or injury from abrasions to the skin. Spaced out amongst the dividing cells are melanocytes which contain pigment, and these cells are sometimes the focus of skin cancers later in life. The pigmented spots or protuberances called moles are clusters of melanocytes, but only rarely do they become cancerous. The pigment granules, melanin, come in many shapes and colours ranging from auburn to deep dark black. It is melanin in hair that gives us our characteristic hair colour, as the hairs themselves contain numerous melanin granules. Grey hairs have few melanin granules in them. In the skin, the melanocytes contain small numbers of melanin granules all bunched up together in the centre of the cell. When we sunbathe, these cells make more granules of pigment and the granules move out to the edges of the cell. The skin becomes brown as the granules are passed across into all epidermal cells. People who have recently become brown by sunbathing appear to get brown again very quickly with more exposure to the sun, since they have already made extra granules in the cells and these simply spread out to protect the skin. The pigments do this by absorbing the rays of the sun and stopping the harmful ultraviolet rays from damaging the sensitive developing cells – a process that can lead to skin cancers. Interestingly we all have the same numbers of melanocytes in our skin, but the more pigmented races have larger, more expanded melanocytes that contain more granules.

Once the delicate stem cells of the skin have divided, they are pushed away from the base towards the surface of the skin. As this happens they become toughened by the inclusion of a protein called keratin, and they make very strong connections with each other. These junctions and proteins give the skin great strength, so that it can be stretched as we bend our joints. Somewhere near the surface of most skin, there is a specialised layer of cells that can no longer divide. As they get pushed away from nutrients below, they become strengthened by an increased number of keratin filaments, and a waterproofing substance is formed between adjacent cells. Nearer the surface, most of these cells of the skin are either dying or dead. Where the skin has to be especially tough, such as on the soles of the foot, the layer of dead and dying epidermal cells can be as thick as one centimetre in people who do not wear shoes. It is generally thinner over the finger tips as thicker skin would stop our hands being sensitive to touch. At the surface, the junctions between the cells loosen, and whole dead cells, now called squames, can be rubbed off as we wash and dry ourselves, or by friction from clothes. As the squames become loose, so they form a small niche that holds bacteria. Then as the squames become detached the bacteria can hitch a lift and travel away from the body. The routine of 'scrubbing up', which surgeons perform before every operation, is carried out to remove these loose squames from the surface of the skin to prevent the squames, and any bacteria on them, from drifting from the surgical staff onto the vulnerable patient.

Because of the way the skin is made, it is quite difficult for antigens and pathogens to enter through the skin unless it is cut or damaged. Should any get through some of the layers of cells, they will be mopped up by special cells called antigen-presenting cells situated in the deeper layers of the skin epithelium. The antigens are collected up into the cells, and are reprocessed there so that they can be given to other immune cells to start off the immune responses that should result in the elimination of the pathogens. The antigen-presenting cells of the skin are called Langerhans's cells. Unfortunately, the Langerhans's cells can also pick up irritants, e.g. harmful chemicals such as components of cement or elements such as nickel, and they become bound to normal proteins from the skin. This combination of irritant and protein is viewed by the immune system as antigenic and cells become sensitised to the complex. The region becomes infiltrated with immune system cells and fluid collects there. This local inflammatory process, which is the root cause of allergic reactions to these substances, is an example of type IV hypersensitivity reactions that were mentioned in Chapter 3. The problem is that once the Langerhans's cells have recognised an allergen, they will react again even more quickly next time they meet it. It can take years without contact with the allergen before the rapid response dies down. Once an allergy develops it is better to avoid the cause of the problem or similar substances. Simply wearing gloves or protective clothing can stop allergens getting to the skin.

When skin is damaged, perhaps burned or cut, the open area may extend deeply into the connective tissues below, so that pathogens can bypass the skin and invade the body. Wounded areas often become puffy with fluid and pathogens survive in this very well. The commonest organism to cause problems in skin is a bacterium called *Staphylococcus aureus*. These bacteria can become resistant to commonly used antibiotics, and will not be killed by them. To overcome the chances of infections spreading, exemplary cleanliness must be observed in places of special risk. In particular this applies to surgical wards, maternity units and burns units in hospitals because the damaged body is more liable to be invaded by bacteria and viruses. In particular, any damaged skin needs to be kept covered to prevent infections. Away from hospitals, certain situations create added risks. Children with eczema have the skin barrier broken, so that *Staphylococcus aureus* and another bacterium called *Streptococcus pyogenes* can get into the skin and cause the infections of impetigo and erysipelas.

Recently there has been an increased awareness of an uncommon infection caused by bacteria that the press have dubbed as 'flesh-eating bacteria'. This condition, more correctly known as necrotising fasciitis, may develop into a serious condition in some hospitalised patients and may even cause their death. The bacterium at the root of this problem is *Streptococcus pyogenes* when it becomes infected by a type of virus called a bacteriophage. The viral bacteriophage lies dormant in the bacterium but can become activated by ultraviolet light and other factors, whereupon it promptly destroys the bacterium and toxic products can cause a form of gangrene in the nearby cells, which kills them. The infection then spreads along the fatty connective tissues of the body. Large areas are damaged, hence the name of 'flesh-eating bacteria' that received so much prominence in the media. This can normally be controlled by antibiotics, but it may be necessary to surgically remove the infected area.

Viruses, yeasts and fungi can also enter the skin especially through cuts and abrasions. Warts are caused by the papilloma virus. Fortunately for us, the body's immune system of the skin normally manages to restrict the spread of virus to just the warty formation, and it is quite safe to have skin warts as long as they are left alone. If necessary, warts with the virus can be removed surgically or frozen off without any problems. If the wart is not completely removed but damaged, the stem cells which proliferate abnormally fast in warts can sometimes spread out and cause a local cancer. The cervix is the region in a woman's body which tightly guards the entrance to the uterus or womb. It forms the start of the birth canal between the vagina and the uterus. These regions often contain pathogens acquired from the male during intercourse. Cervical warts, and those on the penis, are particularly liable to develop in the very delicate and easily damaged skin lining these regions. It is important to have cervical warts surgically removed to prevent the damaged area developing into cancer.

Many women suffer a discharge and itching during certain times of the menstrual cycle which is commonly caused by the fungus *Candida albicans*. Other forms of cystitis can be caused by bacteria. The yeast *Candida albicans* can grow in the skin in moist regions of the body, such as folds of skin, under the female breasts or in the groin where it causes redness, and if the skin is broken other infections can get a hold. Yeasts generally are only a problem if they get right inside the body. AIDS patients often get pneumonia from a small organism called *Pneumocystis carinii*. This is probably a yeast although originally it was described as a small unicellular animal. It gets right into the lungs and causes extreme breathlessness. It is quite difficult to treat AIDS patients with effective drugs for this, as the drugs commonly used can adversely affect the blood-forming cells of the body, and AIDS patients already have very few of these cells due to the HIV (human immunodeficiency virus) infection. Normally, healthy people who take such drugs for the relief of urinary tract infections or typhoid can generally tolerate these drugs since they should have a relatively normal blood cell count.

Even if the skin is penetrated by pathogens, the connective tissue underneath can help to fight infections. Firstly the connective tissue contains cells of the immune system, especially macrophages, that can try to kill invaders. As the fight commences so the macrophages give off chemicals that signal to other cells to come and help. These new cells are normally travelling round the body in the blood, and if there is a high concentration of attractants at any one place they can leave the blood vessels to enter the connective tissues. Secondly, under the skin, in the connective tissues the lymphatics can drain away any tissue fluid that forms when the skin is damaged. With the fluid will be either the pathogens themselves, or small antigenic parts of them that are formed by macrophage or Langerhans' cells. This fluid and its particles are taken away from the skin to lymph nodes where the immune system starts to react to the invaders. Thirdly, the connective tissue itself is made of very large molecules with long wavy arms sticking out each side.

Any pathogens in the connective tissue find it difficult to get through between the molecules, and get trapped there until macrophages can find and destroy them. Fourthly, the connective tissue contains the special set of cells called mast cells (see Chapter 2). These are often activated by allergens or irritants that get past the skin into the connective tissues. When a mosquito bites or a wasp stings, the epidermis is penetrated and the damage extends to the connective tissues below. In the case of the wasp sting, fluid is released by the wasp and this makes mast cells release their granules. A potent chemical released by this reaction is histamine, which relaxes the walls of blood vessels and lymphatics making it easier for fluid and cells to come out of the blood into the connective tissue. As a result, the region swells and feels itchy. Rubbing antihistamine cream into the region helps to prevent further swelling, and the puffiness will gradually disappear.

Ports of Entry

The skin does not present a completely unbroken barrier. All over the body there are sweat glands and hairs, both of which are developed under the skin and penetrate through the epidermis to the surface. Then, of course food and air must enter the body, and waste products need to be excreted. Finally sense organs, such the eyes, nose and ears, are structures that break up the completeness of the skin barrier. Pathogens exploit all of these ports of entry.

Because the secretions from sweat glands are watery, often plentiful and generally flow out of the body, invaders trying to get in by these routes often get swept away. Hair follicles are more susceptible to infection. Although they have sebaceous glands liberating oil for the hairs, the flow is stickier than that of sweat, and less copious. The hair follicles become inhabited by bacteria and even insects and mites. The common follicle mite, which most of us have in the hairs around the eyebrows and chin, appears to do no harm. It merely sits in the hair follicle and lives on cell secretions. Boils, especially in men around the beard area, are usually caused by infections of the bacterium *Staphylococcus aureus*. As pus in the centre of the boil increases, the boil swells up as the pus cannot get through the narrow space alongside the hair. Eventually the centre gets softer and the boil breaks with the loss of the hair at this site. Boils are generally limited to the immediate area however, and do not cause widespread infections.

The mouth is the largest port of entry into the body, and immediately inside the mouth the skin is thinner, softer, not keratinised, and more vulnerable to pathogen damage. However, this delicate membrane is protected by a covering of mucus secreted mainly by the salivary glands. This liquid is 99% water, but the remaining 1% is very important in digestion and as a means of preventing bacteria and other pathogens getting into the body. The digestion of food is started in the mouth with the action of salivary ptyalin. This is a powerful enzyme capable of breaking down sugars in starchy foods such as pasta. Ptyalin can also break down bacteria that have a coating of carbohydrate, as well as some other pathogens. The saliva contains substances similar in action to antibiotics. One of these, lysozyme, can attack microbial cell walls and another, lactoferrin, deprives pathogens of the iron that they generally need as a nutrient. Saliva also contains secretory IgA which is effective in the control of bacteria and viruses. In some diseases, the amount of saliva produced is greatly reduced, so that people with Addison's disease or Cushing's syndrome, or patients prescribed certain drugs that affect salivary secretion rates may suffer more from dental problems and develop an inflamed and sore mouth or a dry mouth, which affects their ability to smell and talk. Other substances in saliva prevent blood clots and yet others, after processing, cause blood vessels to relax. The pain of wasp stings and bites may be reduced by the action of saliva. Thus, licking wounds, as the old wives' tale says, well reduces the chances of infection, increases the blood supply and initiates a fast immune response. Animals usually lick wounds to keep them uninfected; however, sometimes animals must be prevented from overdoing the licking, as the wounds will not heal if the natural secretions of the nearby cells are taken away.

The major airways leading to the lungs are also protected by the secretion of a lot of mucus. This entraps any debris or pathogens that get into the airways. Small, beating cellular extensions, called cilia, move the mucus plus its pathogens up away from the lungs towards the back of the throat from where they can be coughed up or diverted into the gut to be destroyed by the acid of the stomach. Some clever invaders, like the measles and influenza viruses, inactivate the cilia and so can gain entry to the body. However, the innermost air-exchanging regions of the lungs do not normally become infected very easily.

Undigested food and waste products are excreted through the anus, and urine through ducts opening just external to the vagina in women, and at the end of the penis in men. In men, sperm and seminal fluid also leave the body through the same ducts as urine. As discussed earlier, the outside skin is tough and relatively impenetrable but just inside the body, these anal, urinary and reproductive tracts are lined by a mucous membrane rather like that of the inside of the mouth, except that the epithelium is a single-layered structure. Therefore, there is a transition zone as the epithelium changes from external to internal types. This means that the transitional and deeper epithelium is more easily torn and damaged than the multilayered skin on the surface. Such regions are weaknesses in defences that can be penetrated by bacteria and viruses that would otherwise not get through the external skin. It is thought that the virus causing AIDS gets into the body at such sites. Many of the sexually transmitted diseases such as syphilis, gonorrhoea or infections with *Chlamydia* also gain entry at these places. Fortunately, in many cases, the volume of urine and seminal fluid flushes out any invaders and helps to keep the penile ducts uninfected. Also the male's accessory sexual glands that contribute to the semen contain mucosal IgA and other secretions that can kill pathogens.

The vagina is different because, although some mucus does drain out through the vagina from deeper inside the body, there is generally little fluid flow to flush out any pathogens that might gain an entry. Instead the vagina has a multilayered epithelium which infective agents find harder to penetrate than thin skin, and in addition it is tough and so prevents the vaginal skin from getting torn during the frictions of intercourse. Surprisingly, certain bacteria are welcomed in the vagina, and indeed are normally present! Those living in the multilayered soft internal skin control the acidity of the vagina, and so stop harmful pathogens from becoming established there. If the bacteria are killed, perhaps because antibiotics are being taken for infections elsewhere in the body, then the equilibrium is upset and pathogens can become established. This may also occur during a menstrual cycle when the levels of the sex hormones alter the acidity in the vagina. At these times, thrush, which is an infection caused by the yeast *Candida albicans*, may flare up and cause local pain and discharge, only to die down again in the next phase of the menstrual cycle.

In contrast, the eyes, whilst breaking the continuity of the outside skin, are not so much at risk of damage as the previously discussed exits and entries to the body. The eyes are also relatively well protected against infection since the tears, like saliva, contain the natural antibiotic called lysozyme. Tears are produced from glands on the nasal side of the eyes. The fluid is blinked over the surface of each eye, and collected up into ducts which take it back into the cavity of the nose. In addition the lower eyelids have glands that secrete an oily fluid that protects the surface of the eye from drying up. Thus any pathogens or dust that get onto the surface of the eye are rapidly washed away. Acute infections, such as those caused by bacteria introduced perhaps with unsterile contact lenses, usually cause a local immune reaction. Granulocytes, especially neutrophils, are attracted into the area to clear up the infection, but as they increase in number they are lost as a discharge which may cause the eyelids to stick together as it dries overnight. Antibiotics or creams can be prescribed and applied to the outside so that the tears carry them over the eye.

Viruses, particularly those called adenoviruses, can infect the eye and cause a condition called conjunctivitis. The conjunctiva is a fine thin, transparent membrane that covers the eyes and the inside of the eyelids. Infections here often make the whites of the eyes pink as blood vessels underneath become dilated. Viral conjunctivitis will cause a slight discharge and may be accompanied by fever and a sore throat. This is commoner in children than in adults. Bacteria also cause conjunctivitis and one in particular, *Chlamydia trachomatis*, is the most common cause of blindness worldwide. The bacteria are passed by people touching each other, or by flies, after which a chronic or long-term infection results. The severe inflammation distorts the eyelids and the eyelashes cut the eye surfaces. The surface scars and tears cannot lubricate the eyes. Then other bacteria also take hold. Patients with *Chlamydia* can be treated with oral antibiotics such as tetracycline, and eye drops are effective in children and pregnant women who should not have antibiotics unless absolutely necessary.

The cornea itself is a thick clear covering of the eye that imparts good protection. If it becomes invaded with bacteria, fungi or viruses, serious infections can result, leading to blindness and other eye complications. Some bacterial enzymes reduce the thickness of the cornea, so that effective protection of the eye is reduced. Pain and photophobia, which is a desire to keep away from light, are indicators that medical help must be sought. Normally the deeper parts of the eyes are not infected, although after cataract surgery or corneal lens replacements some bacteria do get in and cause infections, although they can be treated by antibiotics. More problematical is when the infections have spread from inside the body to the eyes, as happens in some AIDS patients and those with *Toxoplasma gondii* infections. The whole body must be treated for the infection, not just the eyes.

The nose is protected in completely different ways. There is no complete covering over the entrance, since air could not then get through. Hairs just inside the nose are effective filters for any particles that are breathed in with air, and the nose cells secrete copious amounts of mucus to trap pathogens and particles. This prevents infections from getting into the skin lining the nose and the deeper nasal cavity, and can carry them away from the possible points of entry. Mucus can be got rid of by blowing the nose, or it can drain back into the back of the throat where the normal throat-protection mechanisms can work. If swallowed, the bacteria in the mucus are destroyed in the stomach.

The delicate hearing mechanism of the ears is inside the body, and there is a membrane between the outside world and the noise-sensitive region of the ears to protect them from damage. Wax secreted into the eardrum is also protective. Ear infections do arise, but they usually start from inside the body.

Breaching the Defences

As we have seen, the outer surface of the body and the immediate areas inside are well protected to repel invaders. However, we need to get nutrients and air deep inside us where they can be utilised to our benefit. Thus pathogens can also breach the major defences of the body too. The way in which the lymphoid tissue of the tonsils, adenoids and tongue is arranged in a defensive ring at the back of the throat has already been described in Chapter 2. The leukocytes here are very important in that they monitor the air for the presence of pathogens, and any antigens in food that get past the mouth region. In general, the thin layer of cells forming the skin, and the continuation of this inside the body at places such as the mouth and anus, does not become infiltrated with immune system cells other than at special sites. In contrast, the connective tissues underneath are full of immune system cells that constantly patrol the 'at-risk' areas where breeches in the defences might allow pathogens into the body. These entry points are the airways and lungs, the whole of the gut, the urinary system and the ducts from the ovaries and testes. Any invaders can then be killed on the spot and/or recognition signals (antigens) taken off to the nearest lymph nodes for a more sophisticated immune reaction to be started off (see Chapter 5).

Should pathogens get in with food and avoid immediate destruction, they are rapidly taken to the stomach. The high acid content of the stomach will kill some invaders. To counteract this type of attack, many invaders have a protective coat around them that is resistant to the acid of the stomach. Pathogens will then travel further down the gut. After the stomach, the gut has a short duodenum which is a region of high digestive activity where food is broken down. Pancreatic juices that break up food can also attack pathogens and render them ineffective. Very special protective devices are present here, and these will be described below. Then there is a very long small intestine where the food is absorbed. Because this region is so long, and has an intricately folded surface structure, it presents the greatest area of the body exposed to pathogens. After this is a wider large bowel of caecum and colon which accumulate the waste matter, or faeces, before they are passed out at the anus. Pathogens either get to the small intestine where they cause stomach upsets, or they can gain entry to the inside of the body through any damaged region of the gut wall.

The small intestine of the gut is specially adapted to take all of the digested food and minerals that we need into the body. They gain entry by crossing a single layer of cells. Unfortunately, if food can get across, so can many pathogens. To protect the body from pathogen entry from the gut, mucus and immunoglobulins can be secreted to act on the cell surface, and specialised regions are formed where any antigens that do get in can be processed. Nearly all of the cells of the gut below the oesophagus secrete mucus that can entrap pathogens. Mucus and pathogens are then passed out of the body as part of the faeces. The gut cells also secrete a special type of immunoglobulin which is called secretory IgA. This, like other immunoglobulins, is produced by B cells, but these reside in the connective tissues under the gut wall. The released IgA reaches the surface of epithelial cells facing into the body, and there binds to receptors so that the IgA enters the cells lining the gut wall. Here, inside these cells, two IgA molecules become linked together. This large version of IgA is then secreted out into the lumen of the gut where the food is, and its structure protects it from being destroyed by the powerful enzymes of the intestine that are there to break down food. IgA can bind to, and mop up, any antigens, especially viruses, before they can even enter the body. This form of secretory protection is found on all of the epithelial-lined areas where pathogens may enter: the gut, the respiratory system, the urinogenital system, the mammary glands of the breast, and the eyes. IgA is about 20% of all immunoglobulins produced by the body, so it is a very important protector of the body.

In addition to taking in food, the small intestine of the gut also scrutinises food for pathogens. This occurs in specialised regions called Peyer's patches that are domed out towards the food-containing central passage of the gut. These regions increase in number down the length of the small intestine. The surface of Peyer's patches is covered by flattened cells that do not take in food, but are specialised to recognise and take up antigens. Immediately under the surface, both T and B cells accumulate, and close by are antigen-presenting cells. Thus any antigen taken in with food can be preferentially recognised, processed and presented to B cells so that an immune response can be started against the antigen. In this way, Peyer's patches have a similar function to those of the tonsils and lymph nodes described in Chapter 2 and considered in more detail later in Chapters 5 and 6.

Because gut cells are very active in digesting food, and because they get damaged by the friction of undigested food passing along the gut, the cells of the gut wall need to be replaced very quickly. Not only do the cells become damaged, but larger tears may occur in the gut wall and pathogens will readily get through such cracks. This is where the patrols of immune cells in the connective tissues underneath become really important. All of the small bowel is very well supplied with aggregations of immune cells just under the epithelium. Many of these are the CD8 T cells described in Chapter 2. In some places they form loose collections of lymphocytes amongst the epithelial cells and in the connective tissue of the gut wall, but in other areas they form the Peyer's patches described above. The intra-epithelial lymphocytes bear special receptors on their surfaces that enable them to home to the gut. Once there they reside within the gut wall for most of their life. These cells are predominantly cytotoxic memory cells of the T lineage that can kill infected cells. They do not seem to be able to recognise a wide range of antigens. A small proportion of lymphocytes in the gut wall belong to the γ/δ T cell subset, and these are increased greatly in number in people with cow's milk allergy and coeliac disease, although at the time of writing it is not known how these cells are involved in the allergic responses.

The respiratory system has similar protective devices. The air passages to the lungs also have accumulations of lymphocytes underneath the epithelium of the larger airways. In addition there are many mast cells in the connective tissues, especially around the windpipe. If they are stimulated to release histamine, then the nearby blood vessels become more 'leaky', immune system cells move out of the blood into the area, and the smooth muscle of the windpipe contracts. We have already seen in Chapter 3 how mast cells can be sensitised and create such an allergic reaction. In the respiratory system this is commonly caused by pollen and house mite dust.

Deep in the lungs, the epithelium has to be very, very thin to allow oxygen to be taken out of the air in exchange for carbon dioxide, which is one of the body's waste products. This exchange occurs in blind-ending sacs of the airways called alveoli. If there were many immune cells present at these sites, then gases could not get across from the air to the erythrocytes in the blood vessels inside the body. However, dust and irritants do get very deep into the lungs, so the alveoli there contain macrophages. Although these alveoli are deep in the lungs, the air in them is in communication with the outside world, and the macrophages are technically outside the body! The macrophages move around all the time, gliding over the thin epithelium where the gases are exchanged. They actively take up anything that comes in with the air. Normally when a macrophage is loaded with material it has picked up, it then joins the ciliary staircase and is wafted up out of the lungs to the throat and coughed up and out of the body. The irritant is actually prevented from getting past the epithelium into the body, so there is no need for any of the immune system cells to react to the invader.

If asbestos is breathed into the alveoli of the lungs with the air, then the macrophages try to engulf the particles, which are long, sharp and also totally unbiodegradable. Blue asbestos is more dangerous to the body than the white form, but both damage macrophages in their struggle to take up the particles. The asbestos pierces the macrophage's cell membrane, but asbestos cannot be destroyed by the macrophage, so the macrophage dies and releases all of the phagocytosed asbestos particles back into the alveoli. The asbestos particles can then physically damage the delicate cells and membranes between the air and the blood. Fluid leaks out of the body into the airspace and gas-exchange cannot take place at this site. Asbestos particles become coated with protein and iron to form asbestos bodies. In addition since the macrophages are dead they cannot move out of the alveoli and they accumulate there making the distance between the air and the blood vessels too great for gas exchange. More macrophages are attracted into the region by the local irritation and the cycle continues. At first this local damage does not matter too much since there are millions of alveoli in the lungs but, with time, more and more get damaged and the patient begins to feel breathless. The thin epithelium around the lungs, which is called the pleura, thickens, changes in structure and may become cancerous. There is no happy ending to the story because the particles cannot be removed from the lungs and they go on causing damage. However, the body does try to cope and when the alveoli cannot exchange gases any more, the epithelium gets thicker and the cells start to secrete mucus. This coats the asbestos particles and reduces the damage they do. Unfortunately this is at the expense of the air-exchange surface, so the patient remains breathless.

The general pattern of lung damage described above occurs with many infections too, the differences being mainly that viruses and bacteria can also evoke an immune response and sooner or later immune system cells are involved and antibodies produced. Lung infections can be overcome. Sometimes there is a long-term loss of respiratory surface, but even the alveoli can occasionally be repaired to some extent. It is the unbiodegradable nature of the asbestos particles that makes them so dangerous.

In men, the proximity of the genital and urinary systems and how the flow of semen or urine helps to wash out any infections have already been considered earlier in this chapter. Infections of the urinary system are commoner in women. In women there is only a short distance between the bladder and the outside compared to the similar structure in men, and infections can more easily spread back up to the bladder and kidneys. Cystitis is an infection of the urinary bladder caused by one of several bacteria, usually *Escherichia coli*, *Staphylococcus* species and, in diabetic patients, *Candida albicans*. The bladder becomes irritated by inflammation and the result is an urgent desire to pass water. An antibiotic treatment, plus drinking lots of water or an alkaline drink obtainable from most chemists, helps to flush the bacteria out.

Occasionally infections go even further back into the body to the kidneys – two large organs, each 10–15 centimeters long, that are situated one on each side of the body. They are beautifully designed machines for filtering the blood to remove unwanted substances. This is essential for life, and to have diseased kidneys is very serious. Fortunately, by being right inside the body they are fairly well protected from bacterial, fungal or viral diseases. They do however sometimes have problems because, whilst filtering the blood, they efficiently take out any antigen-antibody complexes that have been formed by the active immune system. Normally this happens without any fuss or bother, but sometimes the immune complexes get trapped in the kidney and in so doing they damage it. Fortunately, even severely diseased kidneys can be replaced surgically although it is important to try to stop the original cause of damage destroying the new implant too.

Subversive Agents

Humans create numerous situations that adversely affect the immune system. The importance of improving hygiene and social living conditions has been recognised for many years. Indeed it has been estimated that more lives have been saved by antiseptic procedures than by the massive expenditure on drugs to combat diseases and control illnesses. The extensive use of antibiotics and vaccines against cholera and tuberculosis alone would not have been enough to control these diseases without the help of improvements in nutrition, sanitation as well as working and living conditions. It has been suggested that the removal of the handle from the water pump at Broad Street in London, by the activist John Snow in 1854, was instrumental in easing the cholera epidemic that raged at that time, although it was already known that aseptic procedures could control the causative *Vibrio cholera* bacterium. This does not underrate the importance of modern techniques in controlling disease. Smallpox would not have been eradicated without world-wide vaccination, but it is essential to recognise that many of our own activities aggravate or allow infection or disease to take hold.

Over the last 30 years there has been a dramatic rise in the numbers of antibiotics available to fight disease. Whilst we have benefited in the short term, we are building up problems for the future. Bacteria have become resistant to many of these formulations by mutations of their chromosomes to avoid being killed, or to produce new molecules that can combat antibiotic actions. In addition to their circular DNA, bacteria have other DNA that exists in the cytoplasm in structures called plasmids. These plasmids can be passed from one bacterium to another across the cell wall, or by the bacteriophage viruses that carry the DNA. The regular chromosomes of plasmids and the bacterium can be altered too by 'jumping genes' that can move genetic material around. By combining genes, bacteria can very quickly develop a good resistance to antibiotics. Today almost all *Staphylococcus aureus* strains throughout the world are resistant to penicillin and ampicillin, and many strains are resistant to methicillin, which was specially formulated to overcome these problems. This is serious since these bacteria are those most often involved in a wide variety of diseases, from superficial boils to deep infections of surgical wounds. Other strains of bacteria in the families of *Streptococci*, *Enterococci* and *Neisseria*, which cause many serious illnesses including meningitis, are still killed by several antibiotics, but the doses needed to overcome them are now much greater than in the past. Improved hygiene is just as important as developing new weapons against these bacteria, and a greater restriction in their use would reduce the phenomenon of resistance to antibiotics and would be of future benefit to mankind.

Sadly many activities that adversely affect the immune system also bring us pleasure, or are socially acceptable customs, so there is considerable resistance by society to deal effectively with these problems. Some activities, such as sex, are only a danger in some unprotected or promiscuous situations. Others are not initially harmful, but carry with them a risk of future damage to the body. Smoking, for example, causes the inhalation of substances damaging to the cilia that waft pathogens and dust out of the lungs. In smokers, these irritants cannot be cleared so easily from the body, therefore infections start more easily and persist longer than in non-smokers. In addition, the smoking habit alters the whole physiology of the body. It changes the levels of several hormones, reduces appetite and nutrient intake, and predisposes cells to become cancerous (see Chapter 9). Chemicals in cigarettes are poisonous to many different types of cell, including those of the immune system. With immunity lowered, cancerous cells cannot be controlled so well in the very early stages so that cancer can become a killer in the relatively young adult.

One of the mixed blessings of recent decades has been the increased use and availability of pesticides and other chemicals to enhance crop production and improve agriculture. Fewer pests mean less damage to food, reduced crop waste and a higher income for the farmer. A higher income means a better standard of living. Nevertheless, the flip-side of the story is that in early years, scientists were not aware of the multiple effects on Man and other animals that some of these create when ingested. Many chemicals are carcinogenic if the correct protective clothes and procedures are not used. Whilst farmers may treat spraying with caution, sheep-dip chemicals, for example, were often splashed over the workers handling sheep.

The rise in the incidence of cancers in the last 100 years is dramatic. In 1899, 30,000 Americans died of cancer, and in 1994 the figure was 538,000. The same pattern of increase is seen in many countries. We now appreciate that the majority of cancers are preventable. The problem is that cell damage occurs long before the cancers are seen. With the increase in living standards seen in many countries, people in general live longer, and may now die of cancer rather than other diseases or infections. Thus past habits and events, such as habitual smoking, exposure to asbestos, over-exposure to sun, accidental or unrecognised ingestion of chemicals from processed foods, pesticides, chemicals or paints, or a myriad of other situations combined with the individual's genetic make-up and life-style, complicate and confuse the issues involved. Even when the cause is identified, and legislation or recommendations devised to limit the damage of cancerous substances, society still has to live with the care and treatment of cancer patients. Our best hope for current sufferers is that the better understanding of cell signalling and control factors during cell division and differentiation will indicate where we can strive to halt, correct or control abnormal cell growth (this is discussed further in Chapter 9).

The hazards of certain self-inflicted life-styles often make head-line news. The high incidence of AIDS in homosexuals and drug abusers, and the lowered immunity of drug addicts are in theory preventable. If the wisdom of older years could be put into young heads, then perhaps drug abusers would understand how they are compromising their immune systems and shortening their lives. Most drugs, especially opiates, have complex actions in the body that affect almost all systems. In addition, those addicted to drugs often have poor diets and life-style, and lack physical fitness, all of which counter the maintenance of good health.

There have been very few in-depth studies on the effects of specific drugs of abuse on the immune system, and more studies are urgently needed. From the small amount of research on the effects of opiates, heroin and cocaine, there is an indication that effects on the immune system are to be expected. Animal models for heroin-use show clear alterations to certain immune parameters, such as the ability of B cells to divide, and alterations in cytokine secretion. These changes could reduce the body's ability to fight infections. Methadone, which is used for treating heroin addiction, has similar but possibly broader effects. Studies in Man are difficult to evaluate correctly because of the problems of keeping track of drug abusers who rapidly alter their life-styles away from normal patterns of behaviour. Additionally, many drug addicts are also HIV-positive, and therefore carry other opportunistic infections as well as the AIDS virus. It then becomes very difficult to distinguish between the effects on the immune system of drugs, HIV and other infections. It is known that in AIDS the infection and loss of CD4 T lymphocytes and macrophages is devastating, and how quickly this occurs is basic to the speed of the progression of AIDS. However other aspects of decreased immune system parameters, such as the loss of thymic tissue and the reduction in T cell numbers, could be caused by any aspect of the AIDS syndrome. The multiplicity of effects resulting from some drugs in the body is seen from numerous studies of alcoholics. In addition to the well-known effects on behaviour which affect all bodily systems, alcohol abuse acts indirectly to immunosuppress patients. This mainly occurs where there is an inadequate intake of protein, vitamins and minerals (see Chapter 8) resulting from missed meals, lack of appetite, diarrhoea and vomiting. There are also direct effects on the gut and cells of the immune system. Chronic alcoholics have a lower acid content in the stomach, and a higher level of gut bacteria compared with non-alcoholics. The cells lining the gut become damaged, so pathogens enter more quickly and poisons from bacteria, such as endotoxin, are more dangerous. This contributes to the alcoholic developing liver damage. One of the effects of this is that antigens cannot be cleared from the body (see Chapter 7). Then there are direct effects on cells of the immune system too. Alcoholics have fewer granulocytes, and those that are in the blood fail to move properly to where they are needed. As a result the innate immune system in chronic alcoholics is poor so that phagocytotic actions against viruses in particular are reduced. In animal models it has also been shown that the kinetics of T lymphocyte production are altered because the thymus, which produces them, shrinks.

Most diseases are not self-inflicted. Some diseases may be induced by accident in situations designed to care and prevent disease. Nurses and doctors looking after patients with infections may have the patient's conditions passed onto them via coughing and sneezing. Dentists, surgeons and emergency workers are at risk of contracting hepatitis and HIV by accidentally being contaminated by blood from infected persons. Such innocent bystanders are infected simply by being too close to a large number of pathogens. Tuberculosis, more commonly called TB, may be a danger to a much wider range of people – in fact anyone in a crowded place close to a carrier. The organism *Mycobacterium tuberculosis* that causes pulmonary TB is common throughout the developing world, but was largely kept under control in developed countries. More recently, with increased travel and the increasing spread of AIDS, there are more cases of TB around. When people are crowded together, especially under poor living conditions or crowded underground trains, the TB mycobacterium can easily be spread from an infected person to others by coughs and sneezes. Anyone with impaired immune responses, such as patients with AIDS, are particularly at risk. Also practices designed to help patients may inadvertently cause disease. Pneumonia occurs more readily in seriously ill patients requiring artificial ventilation of the lungs. The causative bacteria are already in the stomach and lungs but ventilation may assist their movement along the windpipe into the lungs.

How then does one cope with such a wide range of risks? The body provides a good starting point. It is well designed to repel invaders. Keeping oneself in good health means having a good diet, plenty of exercise, avoidance of addictive practices and a cheerful disposition. It may be hard to cultivate the latter, but it does seem to have a bearing on health as will be shown in Chapter 8.

5

The Fight Begins

Meningitis There are several forms of meningitis, caused by viruses and bacteria, all of which result in acute or chronic inflammation of the membranes over the brain. This covering is called the meninges, hence meningitis means inflammation of this region. Many viruses causing common diseases such as mumps and polio, and the enteroviruses can, after infecting the rest of the body, enter the meninges of the brain. Viral meningitis is often slower to develop than bacterial meningitis and is generally milder. Bacterial meningitis is not so common, but is often a much more serious disease which can kill in a short period of time. It may start as an upper respiratory tract infection, fever and a rash, and can be passed on by coughing and sneezing. Once in the body the causative agents pass to the brain via the blood. If the meningitis is caused by *Neisseria meningitidis* (meningococcus) the rash that may occur on the limbs is purplish. This used to be known as spotted fever. *Streptococcus pneumoniae* (pneumococcus) is another infective agent, and in babies it is often *Escherichia coli* or *Listeria monocytogenes* that causes meningitis. In the west of America, in California, Arizona, Texas and Mexico, meningitis can be caused by a fungus, *Coccidioides immitis*. This fungus is very common in the soil and as many as 90% of the population may show evidence of having been infected previously. In severe infections, this form of meningitis may become a killer.

Chronic meningitis is a distinctly different disease, often caused by *Mycobacterium tuberculosis* which causes TB, although many other agents such as *L. monocytogenes*, mentioned above, have been identified. Fever is not present so often, but other neurological complications, such as nerve palsy or paralysis, are more common. All forms of meningitis can leave the patient after recovery with deafness, blindness and nerve palsies although treatments can be given to protect against these post-infective complications.

Any patient suspected of having meningitis is generally given a lumbar puncture to remove a small sample of the fluid around the spinal column for further examination in the laboratory. The fluid of the spinal cord is in continuity with the fluid in the brain and can be examined for the presence of bacteria, leukocytes, protein and a low level of glucose. Lumbar punctures may not be performed where the cause is considered to be viral, as the viruses are very small and difficult to identify although a secondary bacterial infection is sometimes present.

Treatment must be rapid, and the exact chemotherapy will depend on the infective organisms found. Since the family and other contacts are at risk of being infected too, they may also be prescribed medication, and vaccinations for some forms of meningitis can be given, but it seems there are many difficulties in trying to make vaccines to type B of *N. meningitidis*, which is the commonest endemic form of bacteria causing meningitis.

General Defence Policy and the Enemy

Pathogens can come in many shapes and sizes. They can be bacteria, fungi, viruses, smaller protein particles like the Bovine Spongiform Encephalopathy prion that is more commonly called BSE, parasites or even chemicals that trigger immune responses. In very primitive animals, the simplest way to inactivate pathogens is to swallow them and break them down. This process of phagocytosis is the Non-Adaptive Immunity or Innate Immune Response referred to Chapter 1. All higher animals including Man have retained this system as a very important and basic part of our immune system. Furthermore it has become elaborated, and exploited as part of Cell-Mediated Immunity that can be distinguished from the other important method of combating invaders – Humoral Immunity, which involves immune responses relying on the specific action of secretions from immune cells. Indeed 'umor' in humoral actually means liquid. The secretions are antibodies, and they are extremely accurately targeted to interact with particles of invading pathogens. In Man and other animals with a backbone, this system has been developed to a very high degree.

The more primitive system of phagocytosis, compared to cell-mediated and humoral immunity, is much more general. For example, if you cut your finger, tissue is damaged and some cells begin to die. These dying cells release their contents, some of which are harmful to the surrounding cells. But the main effect is to attract macrophages and other phagocytic immune cells to the region which try to clear away the debris. This then allows the remaining cells to create a new environment that will favour the growth of nearby cells and repair the cut. These actions are relatively non-specific. However, if bacteria are accidentally introduced into the cut, phagocytic cells will engulf the bacteria and break them down. Parts of the bacteria will be reprocessed inside the phagocyte, and certain components will be picked out and used to signal to other cells that bacteria have invaded the body. These fragments of the invader are the antigenic particles described in Chapter 1. Such particles are often derived from characteristic components of the invader such as bacterial or fungal cell walls and viral coats. How these particles signal to the immune system and induce a defensive reaction and the cell interactions involved are the subjects of this chapter.

One of the smallest infective agents yet described is the prion. It is not a cell, it does not have a nucleus. The prion is a small protein molecule that accumulates by deranging the normal metabolic processes of the host cell. Prions seem to have a predilection for neural tissues, and the central nervous system in an affected animal or human can contain a very high number of prions. This of course disturbs neural function and death is caused by the effects of neural degeneration. There are about eight known prion diseases, some can be inherited and others are passed from one infected individual to another, often by eating infected offal derived from brains or the nervous system. The diseases of humans called Kuru, which infects some hill tribes in Papua New Guinea, Creutzfeldt-Jacob disease, Gerstmann-Straüssler syndrome, and various related conditions affecting animals – such as scrapie in sheep and bovine spongiform encephalopathy in cattle – are all examples of these slow degenerative diseases of the nervous system caused by prions. Some people think that Alzheimer's disease may also be caused by prions, as there are striking parallels with prion diseases; however, that remains to be proven. Until recently, the incidence of Creutzfeldt-Jacob disease was very low, with only one, usually elderly, person in a million being affected world-wide. With the use of nerve tissue from cows in meat products and in animal feeds, the appearance of Creutzfeldt-Jacob disease in higher numbers and in young people alerted veterinary officers to the possibility of transmission from cows to Man. At present, most people in the UK who have died of the disease are those who, at the end of the 1980s when the disease was at its maximum, either ate a large number of beef burgers, or worked in abattoirs. The only known means of controlling the disease is to kill off all possibly infected stock and to burn the carcasses. Even this has to be done carefully as prions are only killed by very high temperatures.

Much more studied are the viruses. These extremely common forms of life do not multiply on their own either, but need to get into a cell for their proliferation. They are essentially a tiny bag of either DNA or RNA nucleic acids and viral enzymes wrapped in a coat of protein. Some have an envelope around them, derived from the host as the virus leaves one cell for another. Enveloped viruses, of which the HIV, smallpox and hepatitis B viruses are an example, can leave host cells without damaging them. Non-enveloped viruses, such as the polio, hepatitis A and wart viruses, normally cause the host cell to die. Viruses are large enough to be seen by biologists using the magnification powers of an electron microscope, and their protein coats can be distinctive to microbiologists. Thus, although it is difficult to find and identify such small invaders, it is possible to know which type of virus is present in the body. Since the virus needs a cell in which to multiply, viruses have evolved a clever means of targeting the right cells. Most viruses use their cell coats to identify receptors on cells. Viral–receptor relationships can be exploited by microbiologists and pathologists to identify viral infections. In the case of AIDS, the HIV needs to find cells with the CD4 receptor on it, whereas the rabies virus needs a receptor that binds acetylcholine which takes messages between nerves and muscles. The cell with the receptor then becomes the host cell in which the virus replicates. Sometimes the cell is killed by this, as in the case of HIV killing cells with CD4 on them, sometimes the cell tolerates the virus and acts as a carrier, but in other instances the virus inserts its nucleic acids into those of the cell with the result that the cell is dysregulated and is then termed cancerous.

Bacteria are larger, and they can live on their own. Viruses have a cell coat, but bacteria have a cell wall that differs from that of viruses by not being a crystal lattice. The structure of the wall varies between different species of bacteria, and this influences how easily they are attacked by the immune system, and how much damage a bacterium can do to nearby cells. Bacteria differ from all of our cells, except mature erythrocytes, in not having a distinct nucleus. The nucleic acids are mixed with other bacterial components inside the cell. Bacteria are classified according to their shape. Most are spheres called cocci, or rods called bacilli although other shapes such as spiral bacteria do exist. The coccal bacteria include those responsible for meningitis and gonorrhoea, whilst the bacilli include bacteria causing Legionnaires' disease, anthrax, diphtheria, salmonella poisoning, cholera and many gut infections. However, not all bacteria are harmful to us, we need the useful bacteria that exist in our gut to aid our digestion of some foods. Their elimination by antibiotics used against harmful bacteria can itself upset the digestion of food and the natural control of harmful bacteria. This can be a problem for young babies and malnourished adults.

Yeasts and fungi are generally larger than bacteria and, although small, they have cells with nuclei rather more like our cells. This makes it more difficult for the immune system to identify and deal with them. Fortunately the fungi are mainly a nuisance on the nails and skin on the outside of the body, so control is easier. The yeasts often get into the body though, and some are serious causes of disease in the lungs. Much attention is given now to *Pneumocystis carinii*, as it is an opportunist infection afflicting AIDS patients.

The protozoans are single-celled invaders that cause much disease worldwide. Many can only live by spending part of their life cycle in a host other than Man. This has sometimes restricted their spread, so that many are regarded as causing 'tropical' diseases. Most travellers are warned to protect themselves against being bitten by mosquitoes and flies. Malaria, which infects over 270 million people, is carried by several different mosquito species. The family of *Leishmania* parasites causing Chagas' disease in Brazil, cutaneous leishmaniasis and others all live in sandflies, dogs or small rodents as secondary hosts. It is often not possible to control the insect vectors, so visitors to large areas of country may be at risk of catching these diseases. Other protozoa including *Amoeba* and *Giardia* are contracted through drinking contaminated water. In Europe, *Toxoplasma* infections are commonly caught from cats, and by eating undercooked meat, especially beef. In France, where toxoplasmosis is relatively common, about 70% of the population are immune to the disease but in the UK, where it is less common, only about 30% are so protected. Thus UK visitors to France are more at risk than the local population, and getting a *Toxoplasma* infection whilst pregnant can be dangerous for the baby.

Parasitic worms infect Man. These large, multicellular animals are classified as roundworms, tapeworms or flukes, and are not to be confused with garden worms! Fortunately most multiply outside the human body in more primitive animals such as insects or snails, so the amount of infection is directly related to the number of parasitic worms taken in. The roundworms, which are generally small, can be taken in through being bitten by flies, mosquitoes or from food and water. Once in the body they can migrate through tissues and cause damage to many organs. The parasite that causes river blindness is endemic in flies living along the river banks of large tracts of Africa. This is unfortunate, as most communities rely on water for many activities, including transport, irrigation and washing, so millions of people are infected and eventually will be blinded by the *Onchocerca* roundworm. In Australia, south-east Asia and India there are other species of this worm mainly affecting horses, sheep and cattle. In Africa, the Caribbean, the South Pacific Islands, India or Burma, humans can be infected by being bitten by mosquitoes carrying the *Wuchereria* parasitic worms. These tiny worms live in the lymph drainage vessels causing filariasis in over 90 million people. When the parasites lodge in the legs, the legs swell to resemble elephant's legs. Thus this disease is called elephantiasis. Some other parasitic worms such as *Ascaris* get taken in with food or water and live in the intestines. These large parasitic roundworms are one of the commonest of all human parasites.

Tapeworms are also common since they can be acquired by eating infected pork or beef or from being in close association with domestic animals such as cats. These get inside the gut, and attach to the wall by a head which has large hooks on it to enable the tapeworm to hang on strongly. Once in place they can grow extremely long. The parasitic worms called flukes are more widespread though. It has been estimated that some 200 million people in tropical regions are infected with *Schistosoma* parasites that cause the diseases of Schistosomiasis or Bilharzia. This fluke uses a small snail as the intermediary host. The disease is often chronic with serious complications of the liver, bladder and intestines.

With such a range of potential invaders, it is not surprising that the body has evolved sophisticated means of combating them. However, if the parasite does not kill its host, then it will reproduce and survive, and parasites have evolved numerous ways of surviving by circumventing death from the actions of our immune systems. Parasites and humans have evolved together. If either gained the upper hand in the fight, then it would be the end of one of them. Man recently has been winning because of the invention of antibiotics and the increased levels of hygiene and cleanliness used in modern society. Despite this, the rapid rate of reproduction of viruses and bacteria means that their genetic make-up can change quickly and, if an antibiotic-resistant form arises, it will survive to create a new form and reproduce. Parasites that rely on intermediary hosts will die out if one of these hosts is eliminated because the genetic, behavioural and environmental changes needed to adapt to a new host are so complex.

Sentry Duty

For the immune system to act effectively against invaders, it must first recognise that the invader does not belonging to the host's body. This is achieved by the epithelial cells at the invasion site, special sentries posted at danger points for the entry of pathogens, or cells that patrol the whole body to encounter any that have got in. If the invader has attacked the gut, respiratory or urino-genital systems, it may bind to the secretory form of immunoglobulin, IgA, that is on the outside of the cells. This sentry action could halt the infection. The skin is much thicker and tougher than the membranes that secrete IgA so, instead, the skin contains cells specialised to react to antigen – the Langerhans's cells. Although a constant component of skin, Langerhans's cells, once they have taken up antigens, move out of the skin to the nearest lymph node where they can present the antigen to professional immune cells. More mobile though are the lymphocytes, monocytes and granulocytes that circulate continually throughout the body. The blood and lymph systems form interconnected loops whereby mobile cells can travel in the blood to target organs or tissues. Once they have interacted with any antigens present, the mobile cells usually leave the tissues by entering the lymph vessels that drain each organ to lymph nodes. Lymph passes through the lymph nodes to reach larger lymphatics which eventually join the blood vessels. Thus a recirculating pool of cells is able to penetrate most parts of the body to survey all regions, although the brain is a protected site into which lymphocyte entry is carefully controlled and restricted. Only in disease conditions do immune cells commonly get into the brain and nerve tissues of the spinal cord.

Although there is always a recirculating pool of cells, it appears that most lymphocytes have preferred 'homes', and bear specific receptors that help them to locate correctly. For example, some T cells have receptors that specify homing to certain parts of the gut. Once there, these cells stay there to act as sentries. We call these T and B cells 'tissue specific'. We are only just beginning to realise that this homing happens in the normal uninfected state, since most studies on the ways cells move between blood and tissues had been done in the past with regard to acute inflammation. In that situation, cells in the damaged area release factors such as cytokines into the surrounding connective tissues. Two cytokines are especially important in inflammation – interleukin-1 and tumour necrosis factor-alpha. These act as signals to the nearby blood vessels to adapt their structure to let leukocytes pass from the blood to the tissue where they can react to any invaders. This is necessary as the blood, even in small vessels, is pumped along at a very high pressure, so cells need a means of being stopped. The signals act in at least two ways. They cause leukocytes and blood vessel walls to bind to each other, and then give signals to aid changes in the walls that allow leukocytes to slip through to enter the underlying tissues. Inflammation also causes other factors to be released from nearby mast cells that make the blood vessel walls more 'leaky'. Thus immune cells can accumulate at the site of an invasion.

Once a sentry has found an invader, how it is dealt with will depend on the type of cell that found it and how the invader was recognised. The least specific method of identifying invaders is by using receptors on phagocytes, the cells that swallow and kill pathogens. Blood neutrophils are the most important phagocytic cells but eosinophils, natural killer cells, large granular lymphocytes, tissue macrophages and their precursors in the blood, the monocytes, also specialise in phagocytosis. These cells can recognise small invaders like bacteria, fungi and sometimes viruses through relatively non-specific receptors that can bind to parts of the cell walls on bacteria or cell coats of viruses. Factors from the blood known as complement factors (see 'The Weapons of War' below) assist in this process by coating invaders to make them easier to be phagocytosed. The destruction of invaders is particularly effective when this coating process is combined with the recognition of antigenic particles of the pathogen, also through other relatively non-specific receptors termed Fc receptors.

Although viruses can be tackled by the innate immune system, the viral particles are antigenic so the adaptive immune system is also generally activated, and probably will be for most other invaders. The adaptive immune system of cell-mediated or humoral immunity uses sentries with the ability to recognise antigen specifically via immunoglobulin receptors.

Most invaders have identifiable molecules on their surfaces, and very sensitive and specific ways of recognising them have been evolved. As described before in Chapter 2, both T and B cells have antigen receptors on their surfaces. Thus T and B cells in the circulation, or in tissues, can act as sentries. The T cell's ability to react depends on the presence of MHC antigens and is restricted to recognising antigens that have been processed by other cells first. This generally means that T cells do not act as early warning sentries, but are mostly active at lymph nodes and in the spleen. However, in the case of viruses, the viral antigenic components may be present in cells travelling around the body in the blood stream. This is because viruses need to enter into cells in order to reproduce themselves. Here, in the blood T cells act as sentries.

B cells work differently. Almost as though the cell is advertising which antibody it will make, the cells bear part of that antibody on the cell membrane as a surface receptor. Each cell will make antibodies capable of binding to a specific antigen. Thus antigenic particles entering the body or those in the blood can be bound immediately to B cells. Usually this binding also occurs in conjunction with other receptors, rather like the T cell receptors, that together activate the B cells.

Both B and T cells can exist in the body as naive or memory cells. As the term 'naive' implies, such cells are newly formed cells that have not yet been involved in an immune reaction. Memory cells, on the other hand, have and retain some record of this as a memory. Both can act as sentries. In general, the naive cells do not respond to the presence of antigen as quickly as memory cells. In both instances, the antigen will only be recognised if it is on a cell that also has one of the MHC molecules present too. However, the amount of antigen plus MHC molecules needed to cause the naive cell to respond is much higher than in the case of the memory cell. Also, most immune interactions need additional co-stimulatory signals, and again the memory cell reactions are not so dependent on these signals. This means that the memory cells will respond to a smaller amount of antigen, and much faster than naive cells. Nevertheless, there may be far fewer memory cells than naive cells, especially in younger people.

One of the most fascinating areas of current research is how the entry of pathogens in one small part of the body is sensed by nerves allowing messages to be transmitted back to the brain. We have sensory nerves all over the skin and around most organs. When stimulated, these nerves alert the subconscious part of our brain to the onset of an immune response as well as to the generation of pain, such as that caused by a mosquito bite. The brain then responds by altering the permeability of blood vessels where the pain was perceived. However, especially in severe infections of the body, the brain may also have far more widespread effects by altering the pattern of chemicals in the brain to cause sleep and fever. The whole of this story will be further considered in Chapter 7 when the interactions of the immune system and stresses, including the stress of disease, are discussed.

The Alarm Call

As soon as any pathogen or invader has been recognised, the first reactions to it start off an alarm call. Even the simplest response involves several cell types and chemical messengers and a small cascade of events is precipitated. In many cases the first alarm comes from the attacked cell itself which releases chemicals if it is dying. These may attract macrophages or antigen-presenting cells which will probably try to kill the invader and/or process the antigen and then kill cells with antigen in them. This is the work ascribed to the innate immune system. At the same time, the processing of antigen creates fragments that can have antibodies raised to them by the humoral immune system. The early events of the alarm call are mainly cell-mediated immune system actions, with reactions occurring in the blood or at the site of invader entry. Humoral events start in specific organs. Those reactions will be described in 'The Battlefield' section below.

Quite often with viruses, the invader is successfully phagocytosed and killed, without a general alarm call being given out. It is destroyed by a localised reaction at the site of entry. Sometimes we are hardly aware that we have been infected the first time, although most of the many childhood diseases such as mumps and chicken-pox cause strong first reactions. Subsequent infections with the same virus will give another reaction later. The strength of the second reaction will depend on how strongly and efficiently the reaction had been the first time. Viruses such as the chicken-pox *Varicella* virus are not completely killed and persist in the body for long periods of time. They hide in nerves and, since it is difficult for immune system cells to penetrate into neural tissues, they avoid creating an immune response. Many years later, this virus can be reactivated when the immune system is depressed, but instead of getting chicken-pox again shingles develops along the course of nerves. The spots often develop over the forehead along the hair line or in bands around the chest, as the nerves run from the spinal cord at the back, between the ribs, around the chest and towards the front of the body.

Let us suppose, in order to discuss the steps in cell-mediated immune reactions, that the body has been attacked by a virus, perhaps *Varicella* which causes chicken-pox. The virus has to get into a cell to survive. Once in the cell, the cell may release cytokines called interferons that can bind to receptors on nearby cells. This causes the second cell to make anti-viral proteins that help to protect it from being infected. However, if the virus enters a cell, it will multiply there and often kills the cell as it leaves. Dying cells release chemicals that are recognised by other cells as signals of cell death. As a result of these alarm signals, blood vessels dilate, phagocytes and other immune cells enter the region, and small patches in the skin become red and swollen as an inflammation reaction starts. The nervous system may also pick up the signals of cell death and send messages to other parts of the brain and back to the infected site. These messages might alter blood flow to the region and/or cause general whole-body reactions such as alterations to hormone or cytokine release.

If the virus has been engulfed by Langerhans's cells, then the cell begins to process the virus to make antigenic particles to display on its cell surface. During this time the antigen-presenting cells may move away to the nearest lymph node and settle in special parts of the node where they can 'show' the new antigen to T cells (see 'The Battle Field' below). Processing the antigen can take up to an hour, and exactly how it is cut up will influence the type and severity of the immune response that follows. Some ways of cutting up antigen create more effective recognition sites for binding, so the immune response is better. Antigen processing will also affect how an individual will respond to the alarm call and cope with the disease. This is why some people with certain MHC genes are more likely to get one type of disease than other people with another MHC profile. The same disease antigen could be cut up differently by another person and the antigenic fragments, now called epitopes, spark off a slightly different immune response.

In the case of all viruses, including the chicken-pox virus, the viral antigen will be presented to T cells, with MHC class I antigens. These T cells have CD8 molecules on their surface. They are called cytotoxic cells because they can kill cells infected with the presented antigen fragments made in the cell cytoplasm from the virus. The process of antigen presentation is the same as how the cell handles its own proteins. For dealing with these antigenic proteins, the antigen-presenting cells contains specific structures called proteasomes. These can be varied so they can cut an ingested antigen in such a way that it will bind very well to the MHC class I antigens being made at the same time in the same cell. The protein products of the proteasome are moved in the cell to a region where proteins are made – the endoplasmic reticulum. The new MHC class I antigens being made in the cell are held in the wall of the endoplasmic reticulum, where the antigen fragment can be fitted into a groove on the MHC class I antigen. When fitted together, the whole unit is passed into the centre of the endoplasmic reticulum. These are channels inside the cell that allow proteins to move around the cell, but in a defined manner. As the complex moves around, so it can be modified. The complex is moved to a region of the cell called the Golgi apparatus. Here small modifications occur in the complex until it is ready to be placed on the cell surface. The processed antigen, derived from the original chicken-pox virus, then sits in a groove of the MHC class I antigen ready to be presented to class-I-restricted T cells, which are the CD8 T cytotoxic cells of the body. These CD8 T cells will now proliferate to make more of the same type of T cells so that the new cells will be able to kill all cells containing this particular fragment of chicken-pox antigen.

The Weapons of War

The major weapon in humoral immunity is the production of an antibody that acts precisely against only the specific antigen derived from the invader. Thus any population of T and B cells that will attempt to react against the antigen needs to contain some cells with the appropriate receptors for this specific antigen. If the antigen cannot be recognised, then no antibodies will be made to it. We have already seen in Chapter 2 how the genes in T and B cells are rearranged to give this diversity. The number of genes involved and the multiplicative effect of the different possible combinations mean that most of us should have some T or B cells able to react to antigens on an invading organism.

Since the first encounter with antigen involves non-specific recognition by antigen-presenting cells, movement of antigen-bearing cells to a lymphoid organ, cell activation of specific cells, cell proliferation and finally antibody production, it is not surprising that the response takes several days to be effective. Maximum antibody production can take about ten days, and this primary response generally lasts about three weeks. There are several immunoglobulins and each has a different function in the body, so each is secreted at slightly different times and under varying conditions. Which is secreted depends partly on the stage of the immune response, and partly on the microenvironment created around the B cells that secrete immunoglobulins. The family of factors called cytokines are very important here. For example, a high local concentration of interleukin-4 around B cells already making IgG will favour a change to the production of a new type of immunoglobulin such as IgE. Such changes usually result in a more appropriate immunoglobulin being used, and in the case of IgE it will also determine which cells will react because only certain immune system cells will have IgE receptors on their surface. These changing patterns of antibody production are termed switching, and this occurs alongside changes in the genes used by the B cell lines. Memory cells are produced from each of these antibody-producing B cell lines.

Although B cells contain IgD on their surface the first antibodies produced by B cells in an immune response belong to a class known as immunoglobulin M (IgM). These are large molecules, made up of five subunits, which are good at fixing complement (see below). They bind weakly to the antigen, but binding is stronger if the antigen has several binding sites. As the immune response proceeds, large quantities of a better antibody are produced that binds more strongly and effectively to antigen. The main type of antibody, especially in secondary responses to antigens, is IgG. There are several forms of IgG, and each is much smaller than IgM and remains in the blood for much longer. IgG is the main antibody passed to the baby across the placenta or in the milk.

The immunoglobulin IgA, or secretory immunoglobulin, has already been discussed with regard to its secretion across the wall of the gut and other epithelial-lined parts of the body (see Chapter 4). IgE is usually only present at very low levels in the body but, as we have already seen, it plays a very important part in reactions to allergens such as pollen, bee venom and the peanut reactions discussed in Chapter 4. The 'wheal and flare' test, used in clinics to see if the skin reddens when an allergen is scratched on it, is the mild form of this reaction, and hayfever, an extreme form. Some people have higher IgE levels in their blood than others, and susceptibility to allergic reactions runs in families. High levels of IgE also seem to be necessary in the fight against some parasites, although it is not known if there is a specific mechanism that is brought into play, or if the reaction is through the activation of mast cells, as in allergies.

The following account of the structure of these receptors is only given to illustrate where the immunoglobulins actually bind antigen, so any reader interested primarily in the overall picture of immune reactions can skip this paragraph! Although it seems complex, this version is a simplification, and the diagram (Fig 5.1) may make the description clearer. All of the immunoglobulins, whether they are IgG, IgA, IgD or IgE, are composed of two small identical proteinaceous chains of low molecular weight, each of which becomes paired to two longer identical chains of a higher molecular weight. The shorter low molecular weight chains are called the light chains, and the longer ones, the heavy chains. The heavy chain is partly inside the cell, and all the rest is outside. One light and one heavy chain make a pair, and the pairs of chains are folded and pleated to create balls of proteins called globular domains. There are different numbers of domains on each chain, and on different types of immunoglobulin. The domains on the free ends of both pairs of chains are very important, as their components can be varied during cell development to create special binding regions for antigen. These end domains are complexly folded and bonded by sulphur bonds to form a stable region called the variable domain. Within the variable domain, there are six areas, each of which may develop differently. Thus six variable regions form the binding site for antigen, and thereby determine which antibody will be bound to the receptor and how strong the binding will be. Since there are six parts of the variable domain, each of which may differ from the others, very rarely are two antigen binding sites the same. Thus an enormous number of different antigens can be bound, each by a differently structured immunoglobulin. This variation only occurs once when the cell is made and subsequently, when the cell divides again, it will reproduce the same binding site configuration to form a clone of cells with identical binding sites. The other end of the heavy chain, nearer the cell surface, has

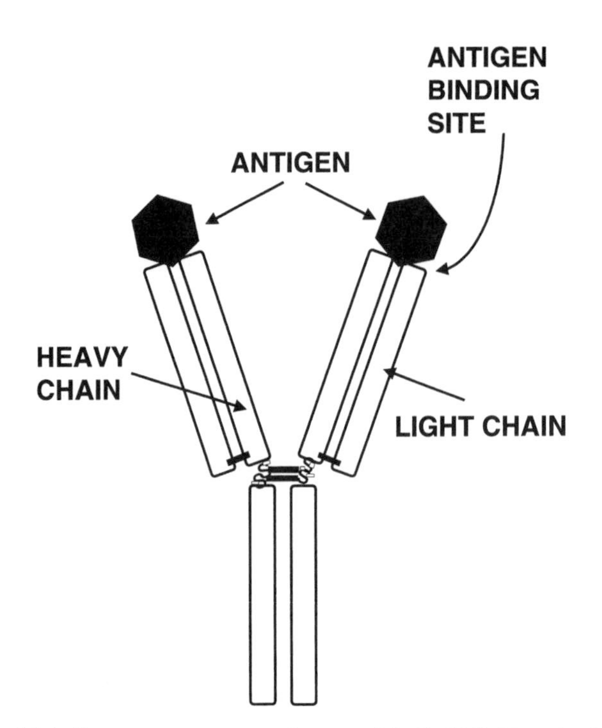

Figure 5.1.A diagram to represent an immunoglobulin G (IgG) molecule composed of paired light and heavy chains. The antigen binding site is composed of three variable domains from each chain which are brought together by the folding of the chains.

constant domains which cannot be varied like the antigen binding site. They do also have a binding function, but this is non-specific.

The weapons, or antibodies, are produced by B cells. When the B cell is first formed, and has not yet met antigen or taken part in an immune response, it is called a naive cell. This does not produce large amounts of antibody yet, although it does have a sample of the type of antibody to be produced secreted onto its surface. Once the cell enters into an immune response, it is said to be activated and will be stimulated into making lots of the same antibody. At first the antibody is generally an IgM but, with time, the type of antibody produced is changed, for example to IgG. These later antibodies are often better at binding the antigen. At this stage, some of the cells no longer enter into the fight by secreting antibodies, but divide to form memory cells. These memory cells can circulate for years in the body, perhaps for the entire life of an individual, and are always ready primed to produce an antibody that will react strongly with antigen against which it was formed. Whether individual cells really can last this long is debatable. However, antigen can be kept in antigen-presenting cells in the centre of lymph nodes for a while, so allowing new memory B cells to be made by existing cells dividing again. Thus memory may live on longer than an individual cell. Because it depends whether naive, activated or memory B cells respond to antigen, the type and speed of responses to one individual antibody are not always the same. The response of memory cells is always the fastest, with peak levels of antibody being released in about three to four days. This is much quicker than the primary immune response by naive cells, which normally takes ten days. In a similar manner, T cells also can be naive, activated or memory cells, so that the whole immune response is speeded up in memory situations, not just immunoglobulin production.

Another important group of weapons in the fight against pathogens is the series of complement factors that influence immune responses and the course of certain diseases such as autoimmunities. Complement proteins comprise about 10% of blood proteins. They work in three main ways by: making pathogens more attractive for phagocytosis; activating immune cells; and causing cells to die. The complement factors are released in a cascade of reactions that is known as the classical pathway, which involves interactions with antibody-antigen complexes (see Chapter 6). In addition, there is an alternative pathway of complement reactions that works independently of antibody-antigen complexes, and is therefore a powerful part of non-adaptive or innate immunity. In both pathways, complement molecules either interact with immune complexes of antibody or directly bind to target cell membranes. Once fixed, the molecules are changed and each new fragment has different actions in immune responses. Some alter vascular permeability, others cause mast cells to release their contents, kill bacteria or other cells, activate neutrophils, or cover bacteria to facilitate phagocytosis. This coating is made of complement substances called opsonins, and this can be recognised by macrophages and neutrophils that have special receptors for opsonins. This action is particularly important in the first phase of an immune response when IgM is the main immunoglobulin being produced. Many immune cells do not have receptors for IgM, so phagocytosis cannot be stimulated by IgM immunoglobulin-antigen complexes. Instead the phagocytes recognise the bacteria because they have been opsonised. In addition another complement pathway can lead to membrane-attack proteins being made on cells that have one of the complement factors stuck there. The attack protein then acts as a focus for the cell-killing complement molecules.

The Battle Field

As with all successful battles, the site of action has to be carefully chosen. The body has predefined the lymph nodes, spleen, Peyer's patches, the appendix, tonsils and adenoids as the main battlefield sites. In addition, loose accumulations of lymphoid tissues along the gut, respiratory system and urino-genital system can also participate in the immune response. There are differences in how the body reacts at each of these sites, but in the main ones there are quite clearly defined principles involved. Firstly the battlefields are structured and orderly, with ranks of cells all positioned in logical order to counteract pathogens. Secondly, there is enormous division of duty with hierarchies of cells specifically adapted to undertake certain jobs. Thirdly, there are generally several means of achieving the same end so that if a pathogen gets past one line of defence, there are other means of dealing with it. Finally, as in several other instances in nature where there is an over-reaction to a response, the amount of antibody produced far exceeds the amount required. This can lead to massive cell death of immunoglobulin-producing cells. Often, thousands of cells must die to enable life to go on, but they are all cleared away neatly so that the infected organs can continue to function properly.

The best battlefield to examine in detail is a lymph node. The body contains lots of lymph nodes, clustered in groups, situated at important sites in the body so that the spread of disease can be limited to regions. Thus the armpits have lymph nodes draining the arms and the chest, the groin's nodes drain the legs and some abdominal organs, and the neck lymph nodes drain the head region. Why is this important? If the bladder gets infected, it is better to have a lymph node nearby to set up an immune defence rather than to allow the infection to go round the body where it might accidentally get into the heart or the brain and do far more damage.

The main activities of the lymph node occur in sequence. Antigen is brought in and presented to immune cells. Then T and B cells interact to create an immune response whereby cells secrete an antibody specific for the antigen presented to the B cells. The secretory cells proliferate, mature and release the antibodies to the blood where they bind with free antigen. Memory cells are also produced for a quicker response to reinfection. Ideally this continues until all of the antigen is inactivated by immunoglobulins, or the cells containing it are destroyed.

Each lymph node is shaped like a kidney-bean (Fig. 5.2). The capsule around the large outer curve has a lymph-filled space inside it called the

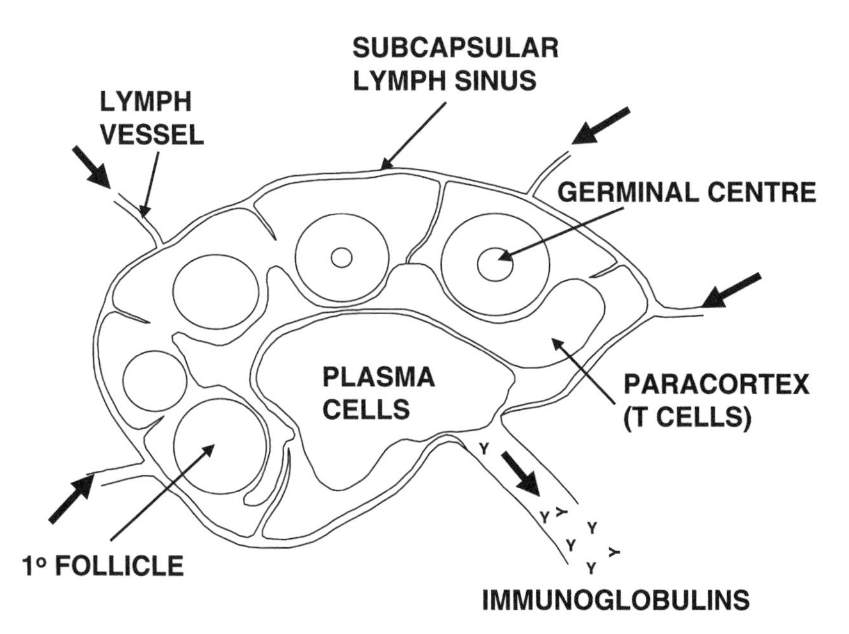

Figure 5.2. A diagram of a lymph node. Lymph can bring in antigens to the node. In an immune response, the primary (1°) follicles enlarge to form germinal centres where B cells are activated. They move to the medulla where they become plasma cells and release immunoglobulins from the node. Help in the reaction is provided by T cells which are located in the paracortical regions.

subcapsular lymph sinus. Numerous lymphatics, which are conduits for lymph from the separate organs, bring lymph into this sinus. The lymphatics do not have a complete wall so they are leaky, and this will allow cells and/or antigens from the draining tissues to get into the lymphatics quite easily. In addition cells may come in from blood. Like all blood vessels, those in the lymph node are always under pressure and are specialised to allow cells to slow down and move out of the blood vessel into the lymph node. The blood vessels assist this process by being made of very tall cells that bear adhesion molecules on the face nearest the blood vessel lumen. Different nodes have different adhesion molecules, so different nodes select different types of cell. Thus the lymph node will accumulate the lymph, antigens and cells from an infected region from either the blood or the lymphatics.

Inside the node there are three main regions (Fig. 5.2). The first is made of clusters of B cells that accumulate together into spheres called primary follicles which increase in size and activity during an immune reaction. Secondly, primary follicles are surrounded by accumulations of T cells in regions called paracortex. Thirdly, the contents of the node including immunoglobulins and cells are drained away on the concave side of the node through the region called the medulla. The immunoglobulin-secreting plasma cells accumulate here during an immune response.

Without immune stimulation the lymph node is small, but it contains all the cells necessary to respond to antigen. These cells are aggregated in, or around, a primary follicle. During an immune response, a reacting lymph node becomes swollen as numerous cells bring in the new antigen and responsive B cells are cloned to expand the population of immunoglobulin-producing cells. As the reactions proceed in the germinal centre in the middle of the primary follicle, it is renamed a secondary follicle. These secondary follicles also swell during an infection and become small again when it is over. This enables a pathologist to examine a sample of the lymph node and assess how the body is fighting an infection. This is particularly effective if the pathologist has markers for the invader. Specific antibodies from a known invader can be made outside the human body using cell culturing techniques. When made, they are labelled on their surfaces, so that when the antibodies are placed on a slice of tissue, the antibodies will bind just as they would have done inside the body. The binding reaction can be made more visible with dyes or fluorescent markers so that by examining the tissue section with a microscope to magnify the cells, the pathologist can see which cell type and how many of them are stained. This tells the pathologist if the lymph node has been infected by the pathogen from which the antibody was made. The techniques can be adapted so that possible cancerous cells in tissues can be identified too.

Antigen can arrive at the lymph nodes from several sources. Antigen in the blood that has escaped being picked up by any immune system cell may filter into the lymph node with the blood or lymph. B cells are especially good at attaching to this free antigen. Earlier (in Chapter 2) we saw how the B cell's receptor for antigen is actually part of the antibody that will react against the antigen. If the receptor and the newly arrived antigen fit together then the B cell will be activated. Activation involves stimulating the B cell to divide to make a clone of identical cells, which are now called plasma cells. The plasma cells from one clone will all make antibody with the same recognition abilities for the antigen, whether they are IgM, IgG or any other immunoglobulin. However, as we have seen, some antibodies bind more effectively than others. Plasma cells take about four to eight days to become fully active and as they develop they move away from the follicles into the medulla. From here the antibodies are released from the plasma cells and pour out into the blood. Some plasma cells leave the lymph node too. The antibodies released will mop up any antigen by combining with it to form large complexes of antibody and antigen. Finally, exhausted plasma cells will die and be cleared away by macrophages. Thus selective cell death must occur for a good immune response.

The lymph node contains various antigen-presenting cells: interdigitating cells that were originally the Langerhans's cells of the skin; follicular dendritic cells; and macrophages. Antigen presentation, as we saw in Chapter 2, has a fail-safe mechanism of ensuring that any circulating antigen from one's own body is not inadvertently reacted against. Reactions can only be mounted against antigen bound to a cell alongside molecules of the MHC system. The MHC molecule may be of class I or class II type. Which it is will determine the type of immune cell that is activated to deal with the foreign antigen, and the type of reaction produced. The most important anti-antigen reactions are elicited when antigen is presented with MHC class II. Cells that can recognise antigen with the MHC class II molecule are T helper or CD4 cells. As we will see briefly below and in Chapter 6, there is more than one type of T helper cell and each secretes different cytokines, so that the type of immune response is tailored to meet the challenge and deal with each kind of pathogens appropriately.

In the parafollicular T cell area of the lymph node, dendritic cells interact with naive T helper cells and present antigen to them. This activates the T cells to secrete their cytokines and to divide to produce more cells of their own kind. Thus a lot of help, of the right kind, is on hand. The B cells in the primary follicles or germinal centres come into contact with follicular dendritic cells that can show antigen to B cells which recognise it with general, not MHC-restricted, signals. All antigen–cell interactions, whether with B or T cells, need other co-stimulatory signals to activate B cells and cause them to divide and then differentiate into fully active immunoglobulin-producing cells. The more precise type of MHC plus antigen interaction of T helper cells also stimulates the same pathway, but this is slower as the antigen has to be taken up by antigen-presenting cells and processed before being placed on the cell surface with the appropriate MHC class II antigen.

The processing of antigen for T helper cells is rather like, but slightly different from, the way that viral antigen or any other cytoplasmic protein is processed by antigen-presenting cells to be fitted into the MHC class I receptor on the cell's surface. This process, which involves internalising the antigen and cutting it up into small antigenic fragments, was described earlier in this chapter. Unlike T cytotoxic cells, which recognise antigen with MHC class I molecules and kill infected cells, T helper cells see antigen only with MHC class II, after which they can act to help B cells make antibodies. As with T cytotoxic cells, new antigen arrives at the cell surface, is taken into the T helper cell, cut into smaller antigenic particles, but then it is fitted into the MHC class II molecule after that has already been made inside the cell. This allows the MHC component to be recycled and used again. The reactions are therefore slightly faster.

Another effect of recognising antigen plus MHC class II antigen is that the T helper cells involved are able to produce and secrete cytokines that can direct the immune response more efficiently. A key player in this part of the story is a cytokine called interleukin-2. This is made by T helper cells of one form that are normally associated with cell-mediated immunity and cell killing. Interleukin-2 stimulates the T cell itself to divide, so making a clone of T cells with the capacity to recognise the original antigen. The other type of activated T cells, called T helper 2 cells, produce a different range of cytokines. These may activate B cells to produce antibody against the original antigen, attract eosinophils into the region, and/or further activate macrophages. Thus these types of fine-tuning of antigen interactions can direct the type of immune response and, if inappropriately activated, misdirect the responses to exacerbate certain diseases.

In the next chapter we shall see how the different types of immune response can be effective and how factors from one response can activate or alter another kind of immune reaction. The differences described above, and many other subtle variations that also exist, may not at first sight seem very different, but the choice of pathways enables different types of antigen to be handled in distinct ways. Thus measles virus antigens will not be processed in the same way as bacterial particles. Although meningitis is the end result of infection by one of several different pathogens, how the body copes with each is different, and the outcome therefore varies not only with the pathogen, but also with the individual's ability to recognise the antigens involved. Additionally, the antigen itself interacts in a variety of manners with different immune cell subsets. T cells need to recognise antigen bound to the surface of antigen-presenting cells in a specific way with the MHC antigen, whereas B cells can recognise antigen that has not been processed. Which kind of cell is able to recognise the antigen, and how it does it, directs the type of immune response the body makes.

6

The Main Campaigns

Malaria Most people will have heard of malaria. It was called Jungle fever in World War II, and Marsh fever by ancient Britons. All travellers are warned to protect themselves if they are visiting a malaria area, such as Africa, Asia, Central America and South America. Other more temperate countries like the United Kingdom can also have malaria outbreaks but these are largely past history. The World Health Organisation has estimated that there are probably 300–500 million cases of malaria a year. This means that it is one of the most prevalent and severe diseases of the tropics. The malaria parasite shares its complex life style with Man and an insect vector, the *Anopheles* mosquito. Mosquitoes can move over large areas of land and take blood from other animals, as well as Man. There are many forms of malaria and several vectors, but all cause cycles of fever and attack red blood cells. Chronic malaria can result in profound anaemia, loss of weight, depression, enlargement of the spleen, muscular weakness and oedema or fluid retention in places such as the ankles. Cerebral malaria is often fatal. Prophylactic medicines protect travellers, but must be started before travelling and continued afterwards. There are no good vaccines at present, although several have been tested. Until these are available, protecting oneself from being bitten by mosquitoes in endemic areas is still the best way to avoid the disease.

Mobilising the Troops

We have already seen how the body's defences are arranged, but sooner or later these defences will be breached. An early and important aspect of invasion by pathogens is the local damage at the site of entry. This signals an alteration to the natural steady state of the body, mobilises immune cells and alerts the whole body to danger. Let us consider a finger that has been cut fairly deeply. It does not matter whether bacteria or viruses have been introduced into the wound by accident, the cut itself is enough to bring in immune system cells. The wound will have damaged blood vessels since tiny ones are present right under the skin epithelium. Tissue injury and cut surfaces release enzymes that activate four main enzyme systems in blood.

These are the clotting system which limits the spread of blood into tissues, the kinin system which signals pain, alters blood flow and vessel permeability, the fibrinolytic system which releases chemicals to attract immune system cells, and the complement system which coats invaders, controls inflammatory reactions and clears antigen-antibody complexes. Not only is each system complex, but all four systems interact with each other at different levels, as well as at different stages of tissue damage and the immune response. Mast cells are central to these responses, as their granules are released by complement factors to act at different stages in the immune response. The granules contain a wide variety of factors, some ready made – such as histamine, chemotactic factors and proteolytic enzymes – and others that cause new substances to be made around the cut region. The ready-made factors tend to increase vascular permeability and blood flow but the newly synthesised ones include factors that cause the blood vessels to contract and help to limit tissue damage.

All mobile immune system cells will be attracted to the region. These will include granulocytes, many naive T and B cells each with a receptor type for different antigens, memory cells with receptors demonstrated to be effective in past encounters with antigen, T cytotoxic and helper cells, natural killer cells, some minor subpopulations of lymphocytes and monocytes that can differentiate into macrophages. Macrophages and neutrophils are general phagocytes, and they will seek out any invaders present. Also, macrophages release a wide variety of signals that attract other cells, activate them, and cause the macrophages to become phagocytic as well as release macrophage factors. In the case of bacteria, but not parasites, it seems that small specific peptides are used to signal their presence and to attract neutrophils, macrophages and their monocyte precursors. The peptides cause neutrophil granules to be mobilised to the cell surface, and also increase the affinities of cell surface receptors to encourage tight binding to nearby cells. Binding occurs when the granules release adhesion molecules which help the neutrophil to locate receptors on the blood vessel walls after slowing down and docking. A series of steps now takes place whereby the neutrophil migrates between the cells of the blood vessel wall to reach the basement membrane under the epithelial cells. This is broken down enzymatically and the neutrophil moves towards areas where attractants are more concentrated to reach the targets for phagocytosis. Some attractants simply signal cell damage, but others are related directly to immune responses. An important attractant for both neutrophils and T cells is interleukin-8 which itself is only released after macrophages secrete cytokines such as interleukin-1 and tumour necrosis factor. These preliminary reactions tend to restrict the infection and its reactions to the local area of a cut. The kind of immune reaction invoked will now depend, at least initially, on the type of invader. It is therefore important to understand how different pathogens are recognised and what type of response they elicit.

Primitive Prions

As seen in Chapter 5, prions are groups of toxic proteins that build up in infected cells. Because they are proteins, our immune system does not appear to be able to recognise them, and they do not invoke any immune response, for example inflammation, although other pathological changes may occur. Unless cells damaged by prions are destroyed by the body by apoptotic mechanisms (see Chapter 9), the best we can hope for is that the non-adaptive immune response of phagocytosis might be able to destroy the cells that contain them, and perhaps reduce or eliminate the prions too. If, in future, prions are all found to be the result of a genetic defect then engineering genetic changes would be a possibility for disease prevention and perhaps control.

Virile Viruses

Most of us, at some time or another, have suffered a bout of influenza. This is caused by breathing in a virus carried by another infected person. We have already seen how the nose secretes mucus to wash the viruses away. If only a low number of virus particles are breathed in, the mucus defence might be enough to prevent infection. With a larger virus load however, the virus will probably get past the nose and into a region less well protected, such as the back of the throat. Those viruses that get into the wind pipe and travel towards the lungs can be trapped on the mucus there and wafted out by the cilia beating towards the outside. Once close to a cell, the virus particles use specific molecules to help them get inside. One of the first results of a cell being infected is that its cilia are prevented from beating. These normally cause a movement of mucus up the respiratory tract by beating towards the outside of the body. Since this does not happen, even more virus particles can get in. Now we can see the medical disadvantages of social customs such as smoking. Smoking reduces ciliary action, so viruses get into the cells of smokers more quickly. Viruses must get into cells to multiply and survive. To aid this, the type-A influenza virus has developed special surface antigens, for example haemagglutinin, as adhesion molecules. These attach to mucosal surfaces such as the nose and the back of the throat, so helping the viruses to get past the mucus barrier which is the first line of defence.

The influenza virus is one of the group of viruses that contain ribonucleic acid and an envelope around the outside that is made from the cell membrane. This envelope has two types of protein in it, both of which are antigenic. Thus antibodies, and therefore vaccines, can be developed to bind to either of these viral molecules. From time to time, the virus makes major changes to these proteins that render the existing antibodies ineffective. This change process is called antigenic shift. This means that the body does not have the right antibodies to fight the new form of influenza, and major epidemics or pandemics arise. These antigenic shifts generally occur periodically every eight to ten years. A major pandemic in 1918–1919 resulted in the deaths of 20–40 million people in the world. Then the virus was described in 1935, and it became possible to study the changes in the virus. Major changes occurred in 1946, and again when Singapore 'flu raged from 1957 to 1967. Another shift occurred in the structure of the virus to produce Hong Kong 'flu in 1968. Smaller changes in the virus are called antigenic drift. Most vaccines incorporate the ability to recognise antigenic change due to drift, but new vaccines have to be developed to combat antigenic shift.

Once the influenza virus has got inside a cell, it multiplies. Some virus infections kill the cell and thousands of virus particles can then be released to infect new cells. This wave of new infections is called secondary viraemia since it causes a rise in body temperature and other 'ill' feelings associated with the immune system's efforts to contain or kill the virus. Other viruses make cells link together without killing them. These viruses can then survive inside cells without the immune system managing to get to them.

The body's defence against viruses depends almost entirely on T cells. One of the most important methods of attack against viruses in cells is a substance called interferon. This is a cytokine produced by activated T helper 1 cells and cells with CD8, as well as by natural killer cells. It causes macrophages to produce cell-killing substances such as nitric oxide synthase and tumour necrosis factor alpha, as well as the cytokine interleukin-1, which has a range of activities in illness, including the ability to raise the body temperature. The cytotoxic substances nitric oxide and tumour necrosis factor alpha come into effect immediately, and help to control the speed of viral replication and spread of viral particles. The longer-term effects of interleukin-1 cause many cells to express more major histocompatibility complex (MHC) class I antigens, and even MHC class II antigens on some cells. This could speed up the recognition of infected cells (see later, 'The Critical Battle'). As was described in Chapter 5, in order to make antigenic peptides for the T cytotoxic cells of the immune system to react to, viruses need to be processed by macrophages or other antigen-presenting cells in association primarily with MHC class antigens. The formation of an antigenic fragment of the virus complexed to MHC class I antigens on the surface of antigen-presenting cells is the first requirement for T cells to act. Any cytotoxic CD8-bearing T cells can now recognise the viral antigen, and themselves become activated. So the range of actions that come into play immediately by the release of interferons leads to an amplification of later anti-viral actions. The response to the invasion is thus multiplied.

Battling Bacteria

Unlike viruses, bacteria can replicate on their own, outside cells. This means that the immune responses to them can be directed to any part of the bacterium, usually to molecules on the bacterial surfaces. As mentioned in the last chapter, bacteria have cell walls. These are of three main types, grouped according to how they can be stained in the laboratory: Gram-positive, Gram-negative and acid-fast bacteria. All of them have an inner cell membrane and a thick outer wall made of large molecules called murein or peptidoglycans. Outside this, the cell wall is very different for the major classes of bacteria, and this affects how the body reacts to the invading bacterium. Bacteria of the first group, for example *Staphylococcus aureus* which commonly causes boils and pimples, have antigenic proteins on top of the peptidoglycans that can be recognised by immune cells. An example of a Gram-negative bacterium is *Salmonella*, which is responsible for some types of food poisoning and typhoid. These bacteria have walls that include a substance called endotoxin which is toxic to the human body. Endotoxin creates widespread blood clotting, changes the secretions of macrophages as well as activating the complement system to generate chemoattractants, adhesion molecules and initiate an immune response. Large doses of endotoxin can result in serious, life-threatening illness. The third group of bacteria contains the mycobacteria, which cause tuberculosis and leprosy. They have a thick outer coat of fatty mycolic acid that is very resistant to being broken down, so they are classed as acid-resistant bacteria. These bacteria often become surrounded by tissue technically called granulomatous lesions that persist and cause inflammation or a type of delayed hypersensitivity reaction.

If the cut on the finger becomes infected with the Gram-positive bacteria, macrophages and neutrophils will try to recognise and then engulf them. However, the recognition of many of these bacteria is hindered because some also have a covering of a substance called hyaluronic acid, which we also have on cells in our bodies. This covers the antigenic peptides normally present, and tricks the body into not reacting against the bacterium. Our immune system has become tolerant to this covering, and cannot react to it. Thus for the immune system's cells to find and recognise the bacterium, this hyaluronic acid coat has to be broken to expose the antigens. Even then, the initial response is a very primitive, non-specific mechanism of recognition that is used against a wide variety of different pathogens. An early effect of this type of recognition is activation of the blood's complement system that can result in the bacterium being coated with factors in the blood called opsonins. Such a process makes phagocytosis of the bacterium by macrophages and neutrophils more efficient. The non-specific mechanisms of recognition include activation of macrophages, T and B cells and other immune system cells that speed up the processes of phagocytosis. Once coated and taken inside the phagocyte, the bacterium will be broken down and reprocessed as described in earlier chapters. Since this process has attracted and activated both T and B cells this part of the immune reaction is more efficient too. The resultant presentation of bacterial antigens in the presence of T cells will spark off B cell responses. When there is no more antigen around, the reactions will die down, perhaps leaving a few memory cells to quickly counteract any other invasions by the same pathogen. The effort of controlling a local infection will mean that some damaged tissue will die, as will exhausted phagocytes. This may form pus in the region of the cut. Such accumulations of pus alter the factors secreted into the local environment, and wound healing may be impaired. However, if all of the invading bacteria are killed then the local immune reaction ceases.

A different type of immune response occurs when the second type of bacteria, the Gram negatives, are the invaders. These are the endotoxin producers, good examples of which are the various *Salmonella* species. *Salmonella enteritidis* can infect both Man and poultry, and handling cooked and uncooked chicken together has resulted in several outbreaks of *Salmonella* poisoning. *Salmonella typhi* on the other hand can only infect Man, so infection must come from person-to-person contact. The cell coats of these endotoxin producers consist of types of sugars with fatty components that can link across to surface membrane antibodies on certain B cells. The reaction is very specific and fairly fast. This may be important since the toxic effects of endotoxin can be fatal. However, this only occurs with a large bacterial invasion, and the numbers in a cut finger are not a danger to life. Endotoxins are released when the bacteria die. The endotoxins create a state of shock, known as septic shock, which causes fever, activation of the complement pathway to control the bacterial invasion, changes in blood clotting, breaks in blood vessels and cell death. Large-scale bacterial infections may also occur after perforated ulcers or peritonitis. In both cases, because the gut wall breaks, undigested food and pathogens can come into direct contact with the inside of the body and cause infection. Clearly the elderly and those with immunodeficiencies such as AIDS or leukaemias are more at risk of endotoxin damage. Patients with extensive burns are also at a high risk of developing septic shock if the wounds become infected.

The toxins released by bacteria can either come from the cell wall, as in the case of *Salmonella* described above, or are exotoxins released by the bacteria. The main medical risks of several bacterial diseases are actually due to the toxins: botulin toxin from *Clostridium botulinum*, diphtheria toxin from *Corynebacterium diphtheriae* and cholera toxin from *Vibrio cholerae*. Most toxins have two subunits, one that binds to cells and one that does the damage. Some toxins kill cells, but others may not be lethal to the cell. Subtle alterations in cellular metabolism may allow the bacteria to go on living successfully inside the body.

Should the bacteria belong to the third group with thicker fatty outer cell walls, then the phagocytes, although able to phagocytose the bacteria, cannot kill them easily. This is particularly true of the bacteria that cause tuberculosis (TB), which can persist for many years in the body and flare up later to cause disease. Sometimes the only defence of the body is to wall-up the macrophage containing the bacteria into structures called granulomas. Granuloma formation effectively compartmentalises the invader by isolating it from the body, and limits, but does not eradicate, the disease. Better for controlling the disease though are occasions when the macrophage containing the bacterium can be helped to kill. Such aid can be provided if there is some degree of bacterial breakdown inside the macrophage. Then, some of the processed antigen will be complexed with MHC class II molecules and can be presented on the macrophage surface to T CD4 helper cells.

This binding of MHC class II plus antigenic fragment to the T helper cell activates the T cell and causes it to proliferate to form clones of cells all able to act against the antigenic fragment. When these activated T cells again meet the antigenic fragment at the same time as MHC class II antigen is presented on the surface of macrophages, the activated T cells immediately release chemical messengers, such as tumour necrosis factor alpha, that activate the macrophage to kill the bacterium inside it. The process can now be repeated and an effective immune response takes place to eliminate the bacteria.

Particular Parasites

Complex parasites are by far the most difficult of all invaders to control effectively. The simplest parasites, as exemplified by *Entamoeba histolytica* which causes amoebic dysentery in many regions of the world, have a relatively simple life cycle. Infection is usually acquired by taking in cysts from food or water. Such cysts are single-celled organisms with a tough protective coat that stops them being destroyed by the digestive enzymes of the gut. Once inside the gut, the parasite penetrates the bowel wall and loses its cyst covering to become a motile trophozoite. In this form it penetrates to the liver, and sometimes to the lungs or brain where it can cause death. This simple organism is hard to detect at the cystic stage, and requires the recognition of different antigens once it becomes a trophozoite. But the more complicated life cycles of many parasites pose even more problems.

Malarial parasites of the *Plasmodium* species, for example, have more complex life cycles. The malaria parasite has been the most widely studied because of its importance to Man. When a person is bitten by an infected mosquito, the parasite enters Man as a merozoite. This merozoite rapidly gets into erythrocytes and multiples there every 70 hours or so, by asexual reproduction. It simply divides – there is no fusion of male and female gametes. This process destroys the erythrocytes and causes fever. After several asexual generations, sexual forms are made but no sexual reproduction takes place inside humans. This occurs in the mosquito, so an infected person if bitten again, allows the parasite to infect the mosquito and breed there. Within the mosquito, the male and female forms of the parasite fuse, and the newly formed stages, now called zygotes, get through the mosquito's gut wall where they form cysts. Inside the cysts, the parasite multiplies and changes shape again to become a sporozoite before migrating to the salivary glands of the mosquito. Now the parasites will be injected into humans as merozoites when the mosquito next takes blood. Each stage of the parasite has its own set of surface molecules that are antigenic, and so too do the cells infected by the parasite. Some antigens are present for a long time, and others for only a short time. Many antigens can be changed slightly in structure so altering the ability of antibodies to recognise them. It is not surprising therefore, that the recognition of parasite antigens in either Man or the mosquito requires different antibodies according to where the parasite is in its life cycle. Also antibodies raised to some antigens will kill well, whereas others are less useful. Adult humans can develop natural resistance to malaria, but only slowly.

The Critical Battle

In considering how the body fights the different types of invader, it should have become clear by now, that there are two major types of reaction – innate or non-specific, and humoral or antigen driven (Figs. 6.1 and 6.2), and a wide range of variations on these themes. Innate reactions primarily kill infected cells or whole pathogens, whereas humoral immunity recognises small fragments of an invader in a wide variety of different guises. The innate, non-specific response is invaluable for a first line of defence, and the best way of coping with viruses and many bacteria.

Although the events that ensue following antigen being recognised in the body primarily involve actions leading to the successful inactivation of foreign molecules, there is no one unique pathway for all reactions. Time and time again, people studying the immune responses of the body see that minor changes to one stage of the immune reaction may have dramatic effects on the outcome of health and disease. Despite this, similar ends can often be achieved by a different series of events. This can be seen in many other facets of medicine. To take a case unrelated to the immune response, we can see that principle in action. Those who have been born with loss of fingers or hands amazingly can learn to use their toes to write, paint and draw. Being blind often heightens sensations felt through hearing or touch. The central immune response of the human body revolves around the unique manner by which the T and B cells create antigen receptor variation (described in Chapter 2) and use this variety of receptors to eliminate antigen. The non-specific methods of pathogen and cell killing are also brought into action to aid in the immune responses.

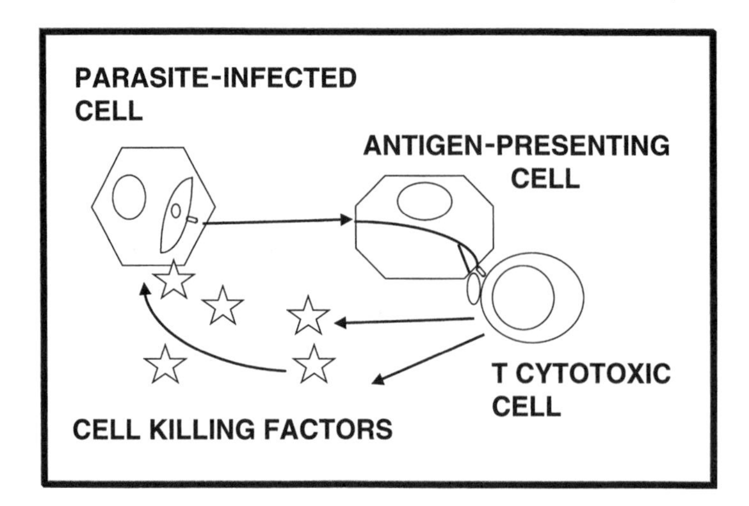

Figure 6.1. The destruction of a parasitised cell can be achieved when the parasite's antigen can be recognised by a CD8 T cytotoxic cell (or natural killer cell).

When something goes wrong, and antigen is not recognised properly, disease is the result.

Innate or non-specific cell killing, as described in the last chapter, is generally performed by macrophage-like cells or specific lymphocytes termed natural killer cells or lymphokine-activated killer (LAK) cells. CD8 T cytotoxic cells need to recognise the target through antigens in conjunction with MHC class I antigens. Not only does this limit mistakes against self, but it also makes the response specific to the processed antigen. The more specific the reaction, the more it can be directed against a particular pathogen. The high degree of sophistication of the antibody-antigen interactions shows this clearly, although it also highlights the major weakness of the system. Change the antigen, even ever so slightly, and the antibody may not recognise it. Thus generally the body utilises both systems against an invader and, at the same time, employs other mechanisms such as the coating of invaders with opsonins or complement factors to facilitate pathogen control.

Within the humoral response pathways, the principal players are B and T lymphocytes, aided and abetted by macrophages and antigen-presenting cells. B cells can interact with any part of antigen, whereas T cells see pathogens as small processed fragments of the original pathogen antigen. The first time a B cell meets antigen, the antigen binds to the B cell's membrane

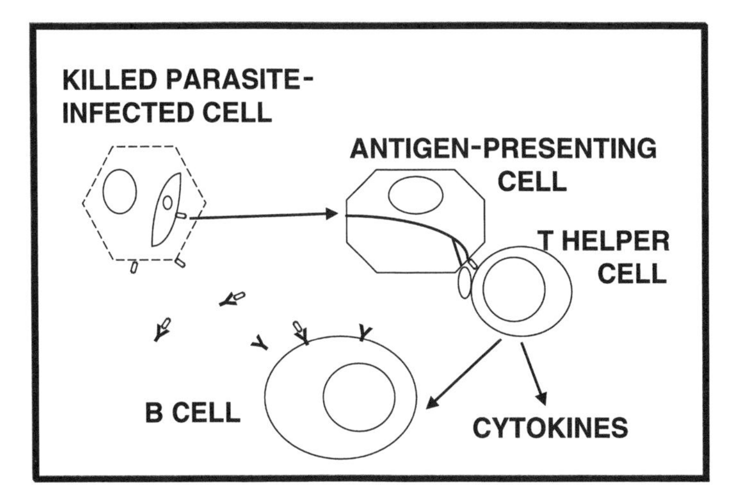

Figure 6.2. The events shown in Fig. 6.1 can also evoke the production of antibodies to mop up parasite antigen in the blood.

receptor which is a sample of the immunoglobulin made by that particular B cell. This membrane immunoglobulin is usually IgM or IgD. The B cell may become activated, depending on the strength of this binding and the presence of other accessory signals delivered by CD4 T helper cells or macrophages. By becoming activated, the B cell can survive, enter the cell cycle to multiply and will secrete new antibody to bind the antigen of the pathogen.

The mechanisms of binding or cross-linking antigen with the B cell's membrane IgM generally involves other signals. One such signal is antigen presented to the B cell in conjunction with MHC class II antigen. Another is interaction between a molecule called CD40 on the B cell, and its binding protein on a T helper cell. These B cells are described as T dependent, and the course of the response will depend on the type of CD4 T helper cell available for interaction (see below). Antigen binding keeps B cells alive by rescuing them from a form of cell death known as apoptosis, and then they can switch to making a new, more specific type of antibody. Which new antibody is made will be influenced by factors around the B cell.

Most surface antigens are recognised by more than one antibody-binding site. Even antigen cut into small fragments and processed inside antigen-presenting cells can bear more than one unique recognition site. Some antigens fit into the antibody's combining site very well, whereas others fit poorly. Furthermore the strength of the antigen-antibody coupling may occur with high or low affinity. The better and stronger the

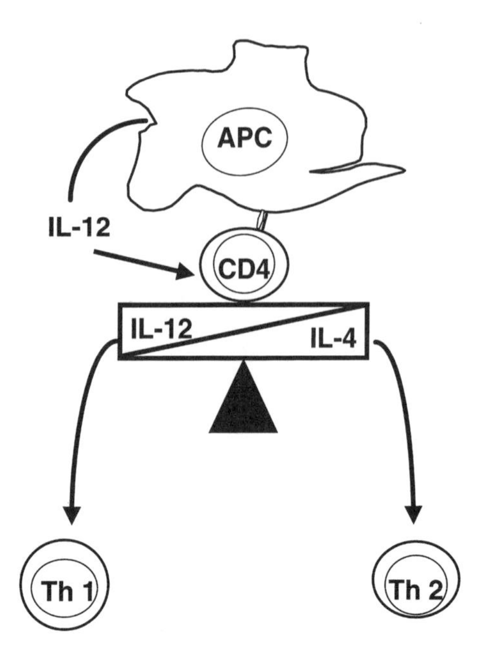

Figure 6.3. Antigen-presenting cells (APC) produce cytokines, some of which (here interleukin-12 or IL-12) are important in determining whether T helper cells become T helper 1 cells (Th 1) or T helper 2 cells (Th 2).

antigen-antibody complex, the better the antigen is inactivated, and the invader checked or destroyed. However, invaders can mutate or change their surface proteins, to create a new antigenic structure. Such new antigen may not be recognised as easily as the previous one, so the pathogen survives longer in the host. Since new receptors are being created on T and B cells all the time, sooner or later a naive cell could be produced with receptors for the new antigen. Thus it is essential that the T and B cells can make a very wide variety of antigen receptors when the body's repertoire is formed, especially in childhood. The activity of the organs where T and B cells are made decreases with increasing age so it becomes harder for new antigens to be effectively dealt with later in life unless they have been met previously.

When naive CD4 T helper cells, which have not been involved in an immune response before, encounter antigen processed by antigen-presenting cells, they change to become one of two types – either T helper 1 or T helper 2 cells (Fig. 6.3). Which is formed is dependent on certain cytokines, especially interleukin-12 which favours T helper 1 cells, and interleukin-4 which favours T helper 2 cells. This change profoundly affects the immune response that ensues, as T helper 1 cells have all the secretions required to produce a good cell-mediated response, and T helper 2 cells favour humoral immunity and the production of antibodies. High levels of interleukin-4 from T helper 2 cells in the vicinity of B cells results in a change of antibody production from IgM to IgE and increases the production of IgG. The release of IgE antibodies can give rise to allergic responses through their cross-linking with allergens on mast cells. However, immune reactions involving T helper 2 cells seem to be fundamental in the control of parasites. Studying T helper 1 or 2 cell function is difficult though as the cytokine interactions are not clearly delineated. For example, T helper 1 cells, in addition to producing cytokines that favour innate immunity, also release interleukin-2 or interleukin-10, which causes B cells to switch to making IgG antibodies. A change in antibody type is necessary for the immune response since B cells that bind IgM and do not receive second signals become non-responsive or anergic, and therefore antigen is not mopped up efficiently. This can create or exacerbate disease.

Some antigens can activate B cells without T cell help. These are called T-independent antigens, such as those found on Gram-negative bacteria with polysaccharide coats. The rapid and specific reaction to antigens of *Salmonella* bacteria has already been discussed. In other T-independent antigen reactions B cell activation can result from cytokine stimulation without antigen being bound to the B cell. However, the antibodies produced are generally IgM, antibody switching does not occur and memory cells may not be produced. So the reaction is not as effective.

Whilst the earliest interactions between naive B cells and antigens can take place in the blood, full antibody production does not. Plasma cells, as fully differentiated antibody-secreting B cells are then called, are most active inside lymph nodes and the spleen. Thus the early events must be moved from the blood to these sites, and the B cells are taken round to them in the blood or by the lymphatics. As we saw in Chapters 2 and 5, the lymph nodes are highly organised. Resting follicles contain antigen-presenting cells and as they accumulate B cells that have interacted with antigen, so germinal centres begin to form. Their formation is antigen driven. The most immature B cells that divide are in the centre, surrounded by recirculating B cells that form a mantle. During an infection, the follicular dendritic cells will contain antigen for presentation to B and T cells. Any B cells that produce high-affinity antibody and interact with these cells receive stimuli to survive, proliferate and turn into antibody-producing plasma cells. If the antibody produced is not of a high enough affinity to interact with antigen-presenting cells, then the developing B cells die.

Cell death itself is not a bad event. Cells must die for the body to live. It is necessary all the time to kill off immune cells that would not make the correct reactions. These may be immature T cells being developed in the thymus that could react against self, B cells with inappropriate antigen receptors, or effete cells that have done their job of protecting the body. Even the death of infected organs may be the best way to clear up the disease since it could allow new healthy cells to divide, differentiate and replace the diseased ones. The body needs cell death to survive. This is fundamental. Most cells will die, unless rescued (Chapter 9).

So far we have seen how the immune system is prompted to react, and how antigens from pathogens are put to use to control the invasion. These reactions must come to an end! It has been shown in experimental conditions that once B cells start to produce a clone of cells producing one antibody, this series of cloning can go on almost without limit. The body does not let this happen. Indeed immune reactions are generally strictly and specifically controlled. Control can be exerted through the antigen itself, antigen-presenting cells, antigen-antibody complexes, effector T or B cells and/or T helper or suppressor cells. Thus the means by which regulation is achieved are many and varied. Many, even major, hypotheses remain to be proven. The antibodies, cellular interactions and alterations in antigen have been studied most. Like the homeostasis mechanisms that regulate and control the numbers of different blood cells, so the antibody itself seems to initiate a regulatory effect. One mechanism that acts at low concentrations of antibody is the binding of antibody to free antigen to reduce the amount of free antigen. Thus only B cells with a high affinity for this antigen will bind to the remaining free antigen. These are then, as we have seen, stimulated into proliferation and the production of that particular antibody. Those that do not bind antigen die. Thus the production of antibody is controlled without compromising the B cells' potential for action. If the concentrations of antibody in the blood are high, another mechanism may come into play. This relies on fragments of antibody binding to antigen and blocking the reactive sites that could combine with B cells. In this instance both the production of antibody and the ability of B cells to be primed are blocked. It is also known that there is a change in antibody binding affinity with time which influences the reactions. The complexes of antigen and antibody themselves may exert regulatory effects. If the complexes contain IgM, the complexes can augment B cell activation, whereas in the case of

IgG, which is produced later in the series of events, the binding tends to be inhibitory. Finally it appears that the antibodies themselves may enable the creation of molecules that bind with the antibody's major reactive site in the variable or V region. This new binding molecule mimics real antigen and could therefore either stimulate or block immune reactions as already discussed. Most of the evidence for such networks is based primarily on experimental data, but it could be responsible for the waves of antibody production seen in some immune responses.

It seems that all forms of antigen-antibody complex can activate the complement system, which is the cascade of blood proteins that aid and regulate some of the immune reactions. This system also helps to remove antibody-antigen complexes as they are mopped up, taken into cells and degraded. This is particularly valuable in the case of viral antigens, because this stops the viruses from attaching to future host cells, and limits the disease. In humans, erythrocytes have receptors for the complement factors that bind the antigen-antibody complexes, so they carry them to the liver and spleen. Both of these organs have large numbers of phagocytic cells that can take up the complexes and degrade them. If the body has been producing a lot of antibody to counteract a heavy infection, then the volume can increase the mass of the organs, and they become swollen. Then organ function changes, and the liver, for example, cannot do its normal job of handling bile and the build up of yellow bile in the body is seen as jaundice. Malaria in particular causes jaundice. With extreme overloading, antigen-antibody complexes end up in the kidney, and there cause other diseases that will be considered in Chapter 7.

PART III – RECOVERY

7

Life after Death

Multiple sclerosis What causes multiple sclerosis is still a mystery, but the strange geographical distribution of its appearance in clusters in certain populations, similar major histocompatibility complex (MHC) genes in many of those affected, more IgG in blood during an attack, autoantibodies to the myelin of the fatty sheath around nerves, and evidence from animals studies of a closely related condition suggest an infectious basis. A common early symptom is inflammation of the nerve to one eye, more often in young women than in children. The disease characteristically flares up in episodes with remissions in between. However, with time, it tends to become relentlessly progressive. The nervous tissue of the brain and spinal column develops hardened regions termed plaques. Here the nerves have lost their fatty protective covering, or sheath, of myelin and there is a strong infiltration of leukocytes into the brain, particularly in the newly formed plaques. This is unusual as brain cells normally restrict the entry of leukocytes by a blood-brain barrier formed from cells within the brain. When this barrier is penetrated, then large numbers of immune cells can enter. Whether these infiltrating cells are making the disease worse or controlling it is unclear. The loss of the myelin sheath results in poor nerve impulse transmission, so that the most devastating practical result of this disease is difficulty in walking. After about 15 years from the onset of the disease, about half of the patients will need a stick for support and can only walk about 50 metres, and some 10% are already wheelchair-bound but can still use their arms.

One or two people per 1000 of the population are affected. Women with certain major histocompatibility class II antigens known as HLA DRw15 and DQw6 (see Chapter 1) have an increased risk of developing multiple sclerosis. As described earlier the type of class II antigen that a person has determines how the body copes with antigen presentation from pathogens. Although the cause of the disease is unknown, this association of particular histocompatibility antigens with multiple sclerosis makes it likely that this disease might have been initiated by a previous infection. However, despite finding that other similar diseases are induced by viruses, and the fact that mice do not develop a comparable disease when they are kept in sterile or germ-free conditions, no virus or infection has been definitely shown to cause multiple sclerosis, even when people in isolated groups of the population have developed the disease. There are higher incidences of the disease at higher latitudes of both the northern and southern hemispheres, with the incidence decreasing markedly towards the equator. In Australia for example, there are five times more multiple sclerosis cases in Tasmania than in Queensland. If teenagers move to a high risk area from a low risk area, then there is a doubling in their risk of contracting the disease. But what it is in the environment that causes this debilitating disease is totally unknown. Within the body, the most strongly implicated candidates are proteins derived from the myelin sheaths themselves. However, the eyes, where multiple sclerosis often starts, do not have nerves with myelin sheaths, so the suggestion is that blood vessel walls may be damaged first. This could then let in T cells, already activated by a pathogen's antigens. If these antigens are in some way similar to myelin proteins, the T cells might perhaps, only under certain conditions, attack myelin sheaths by accident.

These days with the increased use of magnetic resonance imaging in the clinic, the early diagnosis of the disease is better. Thus treatment of acute attacks can be given to hasten recovery and limit damage to the nerve tissue. Specific drugs can reduce pain and help with other effects such as loss of bladder control. Fairly recently it has been shown that some of the body's natural cytokines can be given to successfully reduce the nerve tissue lesions. This is a start, and now other immunologically based treatments are being tried, but they are all still at a very early stage of development.

Dying to Live

In addition to cells dying during or after an immune response, cell death enables controlled growth, the elimination of pathogens and the repair of damaged organs. As the embryo develops, so cells multiply to make the organs and systems of the body. The need for new cells is enormous, and cell divisions proceed rapidly. This vast proliferation in cell numbers must be controlled in a safe way so that each part of the body can function effectively. During the normal development of the nervous system up to 50% of all nerve cells die. To survive, nerve cells must receive signals from the cells they are growing towards. If nerves grow in the wrong direction this does not happen and they cannot live. The body uses a specific and important mechanism called apoptosis for this form of cellular control so that no immune response is invoked. Much of the preceding chapters has dwelt on how antigens create a recognition system for an immune response. How then can cells be killed off without making antigens? The answer lies in degrading cells so they do not work, and then allowing them to be rapidly phagocytosed into small non-antigenic particles that are recycled by phagocytes into nutrients and growth factors.

The first visible sign of apoptosis is reorganisation of the nucleus to break up DNA, and the cell shrinks. Phagocytosis rapidly takes place, even before the cell membrane ruptures, so that there is no leakage of cellular contents into the surroundings and no inflammatory response is evoked. It appears that cells may be able elect to die in this manner if they involve a protein called p53. This protein has been described as a 'guardian of the genome' since it regulates many components of the DNA damage-control process. This was an important finding since the appearance of many cancerous cells is related to deficiencies of p53 in the cell (see Chapter 9).

Cells are doomed to die. Apoptosis is also called programmed cell death. It is now thought that not only is there a limit of about 50 divisions that a cell can make, but the longevity of any particular cell depends on it being rescued from cell death. Once a cell has divided the offspring are programmed to die unless the growth factors, certain sex hormones and minerals such as zinc are present in high enough concentrations to stop the process. Growth factors and messengers called proto-oncogenes give signals that allow the cells to divide, whereas tumour suppressor genes inhibit the action of proto-oncogenes and therefore stop uncontrolled proliferation. The key mediators of apoptosis, however it is induced, are certain enzymes called cysteine proteases, that once activated lead inevitably to cell death. Fascinatingly, many viruses such as *Herpes*, Epstein-Barr and adenoviruses act at this point in the life cycle of the cell and thereby ensure that they are living in a viable cell. By contrast the virus causing AIDS kills cells by attacking the molecular machinery involved in inhibiting cell death.

The ability of the body to let cells die is crucial to immunological events. The very first steps in making effective immune cells relies on apoptosis. Both T and B cells undergo massive gene rearrangements during development (Chapter 2). The elimination of unwanted gene rearrangements, such as those causing the production of potentially autoreactive cells, must be achieved without starting off an immune reaction to the cells being destroyed. Apoptosis is common here. It has been estimated that more than 95% of all developing thymocytes die without leaving the thymus. The reaction is so quick that it is hard to appreciate how prevalent thymocyte cell death is, unless sophisticated techniques are employed. The same is true with B cells in the bone marrow where they are developed.

Once an immune reaction involving the interaction of T, B and antigen-presenting cells has been started, there is a proliferation of cells able to recognise the invader's antigen, to form antigen-specific clones. Nature always overdoes reactions. A clear example of this is the enormous number of eggs and sperm produced and wasted throughout life! Thus as the clones of cells divide rapidly and spread throughout the body to where the new cells can encounter the stimulating antigen, there is a large increase in their numbers. This is necessary. Immune reactions can use the products of a large number of such cells – the more the merrier as fighting the invasion quickly might be life saving. Suddenly though, all the free antigen is mopped up, and the unwanted T and B cells have no job to do. If they do not find antigen, they do not get the signals to go on living and they die by apoptosis. The immune reaction starts to die down. No more free antigen is taken to the lymph nodes, swollen by the increase in cell proliferation. The stimulus to make cells expand in the lymph nodes is missing, and cells die by apoptosis in the lymph nodes rather than divide. The lymph node size returns to normal. Lymph node life quietens down. Similar events occur in parts of the spleen, tonsils, adenoids, Peyer's patches and other mucosal-associated lymphoid tissues.

Cells can also be killed by reactions that involve a binding protein called Fas which also facilitates innate immune responses and allows immune reactions to die down. This is a substance that can bind to any cell bearing a Fas receptor. Crosslinking of cells through Fas results in cell death by apoptosis within hours. T cells have both Fas and its receptor and when CD8 T cytotoxic cells are activated to kill infected target cells they produce more Fas and can bind to infected cells. Both the cytotoxic T cell and its target die. Fas cell killing can also be initiated by Fas binding to antibody on cells. The Fas system also seems to be important in thymocyte and B cell death. It is used to slow down the immune response by deleting activated T cells when they produce Fas on their cell membranes. Another cell-killing pathway uses an enzyme system involving perforin instead of Fas. Perforin literally punches holes in target cells. T cytotoxic cells and natural killer cells only seem to kill cells by these two mechanisms.

Both CD4 T helper 1 and T helper 2 cells also use the Fas pathways for controlling cell proliferation. Indeed the T helper 1 cells (involved in cell-mediated immunity) use it more than T helper 2 cells which help B cells (humoral immunity). The T helper 1 cell produces interferon-γ and tumour necrosis factor, which causes B cells to produce more Fas and die, so preventing B cells from accumulating. Too many might cause problems by allowing autoimmune diseases to develop (see Chapter 9).

It has now also been appreciated that overactivity of the Fas system may be a part of the disease process in some illnesses, especially HIV infections. It seems that there is a specific protein on the envelope of the AIDS virus that can cross-link to CD4 T lymphocytes. This results in a lot of Fas being present on the surface of the lymphocytes. Very high levels of Fas have been found in HIV-infected children and it is suggested that this may cause the death of lymphocytes. Again the T helper 1 and T helper 2 story is relevant. In one study of patients with HIV who did not show any symptoms of disease, about half had good immune responses against other infections such as influenza. These people had produced strong responses associated with T helper 1 cell action. Slowly with time, as the disease progressed, their responses shifted from T helper 1 to T helper 2 reactions. It seems that T helper 2 cells cannot stop virus replication as effectively as T helper 1 cells, which produce the antiviral substance interferon-γ. Further work is currently in progress to try to establish clearly if a change from one type of T helper cell to another is important in the control of diseases. If so, such findings might form the basis for strategies to be used against HIV disease progression.

There is another important way in which a cell may die. This is called necrosis and is a process that ensures an immune reaction is sparked off. When some parasites invade cells, or some viruses leave cells, the cell membranes are broken, the contents pour out and cell death by necrosis ensues. The products from the cell's destruction contain factors that attract phagocytes, and stimulate the innate immune reaction. This is called inflammation. Macrophages and other phagocytic cells quickly arrive and engulf the dying cell. The liberated cell contents also affect factors in the blood and start clotting and complement reactions that make the blood slow up in the environment and make the fragments more attractive to phagocytes. Necrosis can also be caused in organs by the toxic by-products of infections. An example of this type of reaction is given in Chapter 6, which describes the toxic effects of endotoxin. Some poisons and dangerous chemicals can also have a similar effect by direct action on the cell.

Fiery Fevers

The immune responses of the body involve an enormous amount of cellular activity – the production of receptors and factors to start the response, numerous new cells and copious amounts of secretions. All of this takes energy. It is not surprising that the infected person feels exhausted and wants to sleep. At the same time the body's temperature may rise, and then there is no desire for food. What induces this fever, and is it good or bad for the immune response?

Body temperature varies throughout the day around a thermoregulatory set point. Thus four categories of body temperature were defined in 1968, and these still supply a useful basis on which to work. Normothermia is the state when the actual body temperature and the thermoregulatory set point coincide. This is the situation most of the time. In a fever, the set point is raised but the body temperature may or may not be at that level. The two other conditions of hypo- and hyperthermia occur when the actual body temperature is below (hypo) or above (hyper) the set point, whatever level it is set at. Thus, when the thermoregulatory set point has been set higher than normal, an infected person has a rising temperature, and they actually feel hypothermic. This results in a variety of heat-conserving and heat-generating physiological reactions. The patient feels cold, shivers and wraps up. Once the temperature reaches the set point, the patient feels happier. If the temperature continues to rise above the set point, when the temperature is said to 'break', the patient throws off the bed covers and tries to lose heat.

The body's response to infections follows a well-characterised course that is called the acute phase response. Pathogenic substances that initiate the acute phase response include bacterial products such as muramyl peptides, endotoxin or lipopolysaccharide, and viral double-stranded RNA. During an acute phase response, cytokines are released from many immune cells such as macrophages, and acute phase proteins from the liver. The result is an increase in the number of neutrophils in the blood, the presence of acute phase proteins in the blood, a rise in temperature, a desire to sleep and a change in behaviour patterns. The major players in this are the cytokines, the mediators of short-range signals between cells. Some of these are called endogenous pyrogens or substances that cause fever, and others are endogenous cryogens, which attenuate fever. One of the first cytokines to be recognised as an endogenous pyrogen was interleukin-1, which exists in an α and a β form. However, it now appears that tumour necrosis factor is probably the most important instigator of fever, whilst the interferons (α and γ) and interleukin-6 also play a part. Fevers probably die down when certain cytokines such as interleukins-1 and -2 stimulate the tiny pituitary gland above the roof of the mouth to release the building blocks of a hormone called α-melanocyte stimulating hormone. This and another hormone, arginine vasopressin, together cause the fever to die down.

Interleukin-1 is also released from many immune system cells such as macrophages, T and B cells as well as other cells in the brain and those lining blood vessels. Thus it is not clear if any action of interleukin-1 in resetting the thermoregulatory set point is due to circulating interleukin-1 getting into the brain, or whether it is produced locally in the brain. The interleukins, either alone or by acting through such cytokines as interleukin-6 or prostaglandins, cause sleep, fever, the release of acute phase proteins from the liver, the breakdown of muscle and bone, and the movement of neutrophils into the blood. The prostaglandins have a wide range of actions, but in immune responses they help to make the responses of T and B cells, macrophages and neutrophils stronger. The endogenous pyrogen interleukin-6 acts preferentially on B cells causing immunoglobulin secretion. It is one of the most important mediators of the acute phase response and, by acting on the cells of the liver, factors are produced that are needed for the complement cascade which, amongst other actions, coats pathogens for better phagocytosis.

Thus many immune system factors are involved in fever, some having a direct influence whilst others probably work through a cascade of reactions. Overall it appears that their actions promote a better immune response, and since the immune reactions are all energy demanding it is not surprising that fever is accompanied by reduced physical activity and the release of nutrients for re-use in the body. Such a situation would make one weaker. Perhaps it is not surprising that fever also induces sleep.

Soothing Sleep

A raised temperature induces sleep even if the normal complement of sleep has already been attained. Sleep is usually divided into two main states, slow-wave sleep (SWS) and rapid-eye-movement sleep (REMS). REMS is also called dream sleep. There is a continuous cycle of waking, SWS and REMS which in Man is about 90 minutes long, so that four or five episodes of cycles occur during one night. Exactly what induces sleep in the healthy body and where sleep is controlled from in the brain are not really understood. It is known though that there are sleep factors in the blood, and that when these are given to animals they cause deep sleep. The sleep factors have two amino acids only found elsewhere in bacteria. We all have bacteria in our gut, and macrophages constantly break them down. One of the products of this breakdown is a small protein called muramyl dipeptide. Babies do not have gut bacteria for about a month, after which they are able to sleep deeply. Furthermore antibiotics that kill off gut bacteria lower the body temperature and alter sleep patterns. For these and other reasons, it has been proposed that we get our sleep factors from the breakdown of gut bacteria and that these, rather like vitamins, are essential and cannot be made by us.

Gut bacterial sleep factors are not the whole story though. Bacteria, as we saw in Chapter 6, can produce endotoxin. Some of the components of endotoxin, such as lipopolysaccharides, cause the release of many cytokines and other factors that can induce sleep. So too do fungi, viruses and parasites although how important this is has not been determined. Even in sleeping sickness, in which sleep is induced (although not as much as the name suggests), the contributions made by pathogen chemicals and the normal immune response's active molecules have not been clearly delineated. However, viruses do have special sleep-inducing mechanisms. Sleepiness and fatigue are characteristic of many viral conditions especially influenza, infectious mononucleosis caused by the Epstein-Barr virus, and post-viral fatigue syndrome. It appears that RNA viruses especially can make double-stranded RNA, perhaps by mixing RNA with that of the host during the course of infection. This double-stranded RNA induces fever and sleep and other pathological changes associated with viral infections. The RNA, muramyl dipeptide and lipopolysaccharide from endotoxin all use the same type of binding and signalling receptor in the host cells, and therefore all stimulate the immune system to release sleep-inducing factors. All three result in a rise in the amount of interferon-γ produced by T lymphocytes, and we have just seen in the last section that this is associated with fever.

Tumour necrosis factor is somehow associated with deep sleep, and it has been suggested that disturbances in the normal release of tumour necrosis factor may be responsible for the drowsiness and weariness of those suffering from cancer, AIDS and chronic fatigue syndrome. Tumour necrosis factor is produced by macrophages, T and B cells, natural killer cells, brain cells called astrocytes and cells in the liver. The range of activities enhanced by tumour necrosis factor is very wide. It is a powerful modulator of many parts of the immune response, as it induces the appearance of adhesion molecules on cells, activates neutrophils and causes the release of many cytokines from other cells. It was originally called cachectin because it was discovered in studies of the state of wasting and ill health in disease, which is called cachexia. The action of tumour necrosis factor against tumours such as cancer of the pigment-forming cells of the skin, or malignant melanoma, has been exploited successfully in the clinic for some patients.

Chronic insomnia may have many causes, but it is often associated with depression and susceptibility to infections. Depression and insomnia are associated, in many cases, with low levels of a brain neurotransmitter called serotonin. The seasonally affective disorders, for example the SAD syndrome which was discussed in Chapter 3, are related to low levels of melatonin. Serotonin is a building block of melatonin and both need ample supplies of the precursor tryptophan. Eating turkey and drinking milk both supply high levels of tryptophan, and it is interesting to note that a common suggestion for relieving insomnia is to take a glass of warm milk before retiring for the night. Some natural remedies for insomnia like chamomile, valerian and passion flower may contain natural sleep-inducing factors.

Several studies have shown that sleep is beneficial to good health, and affects the immune response. For many years, a standard response to illness was bed-rest and sleep. It is only with a modern fast life style that such 'luxuries' are ignored. Are we right though? It would appear we are very wrong. There is ample evidence that a shortage of sleep, often described as feeling 'run-down', predisposes to infection and illness. Halving the natural sleep requirement also halves the number of natural killer cells in the body. This is reversed after sleep. Thus immune surveillance is probably compromised when sleep is short. Total sleep deprivation in animals leads quickly to death, often from normally harmless pathogens that are always present in their environment. Stress and sleep deprivation together are particularly potent in lowering resistance to disease. Some animals studies do not confirm these findings and there are always some fit, perhaps hyperactive, people, with good stress-control abilities, who appear to defy this, but the majority of people find it hard to resist infection when stress and lack of sleep occur together.

Repairing the Damage

The first requirement of the body after mounting a major immune response is to clear up the debris from dead cells and immune reactions. Antibody complexed with antigen on cells will result in the cells being killed off and the debris phagocytosed. This is very economical since once the antigen is inactivated by immunoglobulin there is no further risk of cellular damage, and the entire contents of the infected cell are broken down and recycled for further use. Worn out macrophages or those laden with undigested products now slowly make their way from all parts of the body to the lungs. From here they get out of the body into the lung's air space, and are moved up the ciliary stairway out of the lungs to be coughed up with sputum and expelled from of the body.

More commonly however, antibody-antigen complexes form in the blood. The complement system of proteins in the blood work on these complexes to keep them soluble so they can be disposed of. However, if they persist or are deposited in the joints, etc. they can be the cause of some, the type III, hypersensitivity diseases, and will require treatment. Most complement-treated antibody-antigen complexes are taken up by erythrocytes that recognise and bind specific complement proteins to receptors on erythrocyte surfaces. The erythrocytes rapidly move around the body and reach the liver. Here the complexes are broken apart and special macrophages in the liver phagocytose them. Eventually these special macrophages move away from the liver to join the exodus of all macrophages to the lungs, from where they are coughed up out of the body in sputum.

The complexes are not all formed in the same ratio of antigen to immunoglobulin, and these ratios vary during an infection. Initially in the blood there is much more antigen, then equal amounts of antigen and antibody, and lastly more free immunoglobulin. Also, during an infection the type of immunoglobulin fighting the antigen changes from IgM to other forms such as IgG (see Chapters 5 and 6) so the binding ratios change all the time. This affects the size of the complexes. Large complexes are taken up within a few hours by the liver and smaller ones circulate for several days. Sometimes immunoglobulins bind less well to antigen so there is an excess of antigen in the body for a long time. Another way that antigen excess can exist is when the immune response is poor and not enough immunoglobulin is produced to combat the antigen. In these situations of excess antigen, the complexes may be poorly coated with complement and therefore not taken up so well by the liver. They circulate longer in the blood and soon find their way into the kidney. The kidney is specialised to use a high-pressure blood system for sieving unwanted molecules out of the blood and excreting them. These complexes are forced to the membranes where filtration occurs. Smaller complexes may get through, but the larger complexes get stuck there. This clogs up the filtration membranes, and disturbs the normal kidney functions. Some of the kidney's cells proliferate and distort the structure of the filtration regions. As the cells in the kidney become damaged so leukocytes are attracted to the region to fight the 'infection'. The affected person has raised blood pressure, excretes protein in the urine and has an inflamed kidney. Kidney failure can result. This is life threatening. The name for this condition is glomerulonephritis, but this term also covers a related immune reaction in which antibodies are formed that act to destroy the kidney's filtration basement membrane. Again kidney failure may result, but this condition is not the direct result of an infection. In children, glomerulonephritis is most often caused by certain Streptococcal bacteria. After the primary immune response, which normally takes about 10–12 days, the face swells up, there is high blood pressure and the excretion of a reddish urine which, laboratory examination reveals, contains protein. This is unusual since normal urine does not contain any protein. This form of kidney damage normally resolves and no further problems arise, although in adults it may become chronic and can cause kidney failure. Sometimes this type of kidney damage is fatal. In an autoimmune disease called SLE, which is short for systemic lupus erythematosus, almost all of the organs of the body are attacked and destroyed by the body's own immune system. This results finally in the kidneys being overloaded with immune complexes, and the patient may die of kidney failure.

The site of immunoglobulin-antigen complex deposition is not always the kidney. Blood pressure and flow both influence complex deposition, so complexes may be deposited where arteries branch in the body, or where blood is filtered before it enters the brain, or in specialised parts of the eyes, for example the ciliary body. The type of disease may also influence the site damaged; for example in rheumatoid arthritis, complexes accumulate in the joints causing pain and stiffness.

As an infection is successfully fought, so fevers die down, less sleep is required and food begins to become more desirable. There is still a need to replace lost fluids, and highly nutritious food is important since dietary function may have been altered and the regular intake of vitamins disturbed. This is particularly true for the water-soluble vitamins such as the vitamin B complex and vitamin C that we must eat each day since we cannot store them in the liver. During recovery appetite returns, especially if stimulated by fresh air and exercise, and the body can rebuild its strength fairly quickly.

There are many instances where this recovery is slow and may even take years. Patients recovering from many infections, whether they are bacterial tuberculosis, parasite-induced diseases such as toxoplasmosis, malaria, or giardiasis, acute viral infections such as influenza, hepatitis A or the common cold, or chronic viral infections from the Epstein-Barr virus or enteroviruses, all show fatigue. In all of these cases, the fatigue occurs in most sufferers, and is described as part of the disease. However, in the summer of 1934, after an outbreak of poliomyelitis, almost 200 medical and nursing staff in Los Angeles suffered from what was termed an epidemic of post-viral fatigue syndrome. Since then many other outbreaks have occurred in many countries including Switzerland, UK, Sweden, South

Africa and Australia, often associated with a poliomyelitis-like infection, and involving large numbers of people. The way it spread in the communities pointed to an infectious disease, but none was identified. Indeed, post-viral fatigue syndrome is denied by some members of the medical profession, written off as mass hysteria or ascribed to depressive conditions by others, yet accepted and described by neurologists and virologists. It is an identifiable condition of immense fatigue with abnormalities in muscular function and it usually follows insidiously from an infection often caused by viruses. Apart from the poliomyelitis-like symptoms, a number of different viruses have been associated with the disease, but none has emerged as clearly causative. An infectious basis for the outbreak in nearly 150 staff at the Hospital for Sick Children, Great Ormond Street, London was substantiated by clinical confirmation of cervical lymph node involvement, mild neurological conditions and a few indications of pathogens. However, no infective agent could be found. Like most other post-viral fatigue syndrome sufferers, most patients appeared to recover after about two years. The condition today, sometimes called myalgic encephalomyelitis or ME, is often applied to chronic lassitude in individuals who, prior to an infection, were healthy, outgoing, non-depressive types. However, controversy still abounds as to the reality and possible causes of this condition.

Rather similar behavioural reactions can occur after poliomyelitis. Despite the widespread availability of good vaccines, some 250,000 people world-wide still get poliomyelitis each year. The vast majority, around 90%, will have no more than mild symptoms including fever, headaches and a sore throat. However, another 10% will have severe reactions which may include meningitis and pains in the neck and back, and a very few of these, perhaps less than 1% of all sufferers, may be permanently paralysed. It has now been realised that in some patients who had suffered muscle paralysis, a fatigue syndrome persisted long after the acute disease had passed. This is now recognised in the USA as PPS or post-polio syndrome, and a network of clinics has been set up to diagnose and help those afflicted. The sufferer experiences fatigue, new muscle weaknesses, difficulties in sleeping and emotional distress that share similarities with post-viral fatigue syndrome already described. Although PPS seems to have a slow onset and muscle loss increases, it is not life threatening. However, of the 1.4 million Americans who have had poliomyelitis, it is thought that about a quarter have already developed PPS, and more may do so in time. The reasons for such long-term effects are not at all clear although the long persistence of viral fragments has been suggested as a cause of the problems.

Lasting Relationships

The commonest forms of the long-term effects of an infection are when the pathogen is not totally eliminated but manages to continue to exist in the body. This is often the case with viral infections, and almost all of the parasites can be very difficult to eradicate. In some cases the individual overcomes much of the infection and manages to live with the pathogen, but in other cases chronic disease and early death may result.

Any antigen that gets into the body can be reacted to or ignored. A non-responsive reaction can be due to either the antigen not being recognised or antigen recognition not evoking a response from immune cells. Bacteria maybe non-antigenic by having an acid-resistant covering over them (see Chapter 6), or by immediately changing the nature of the antigen first recognised. However, sometimes, when the antigen had been recognised and reacted against in the past, no immune response is elicited when the body meets the antigen again. This lack of response to a repeat challenge is called tolerance. There are several ways in which this may arise. In some cases whole clones of potentially reactive lymphocytes are destroyed in the body, or lymphocytes may be rendered anergic or non-responsive, or new lymphocytes may be formed that are able to suppress the reaction. The ways in which tolerance may be involved in disease will be briefly discussed here and in Chapter 9.

As we have seen, the body is very well designed to identify invaders, but should the barrier be penetrated then a myriad of different reactions can be set in motion to destroy the invader and return the body to the pre-infection state. This does not always happen. Since it is in the pathogen's own interest to survive, many ways have evolved to avoid immune detection.

Viruses may do this using several different mechanisms. We have already seen, in Chapter 6, how rapidly the proteins in the cell coat can be changed to elude identification by any existing antibodies, and they can also shift their genetic make-up in larger jumps to produce new forms of disease. This has been clearly demonstrated in the case of the influenza virus when such shifts cause new epidemics. Also, viruses can exist in cells without causing disease when they are said to be in a 'carrier' state. Others can become silent, or exist in a state of latency by not replicating. The most well known example of this is the *Herpes simplex* virus which, after infecting the skin and other epithelial surfaces, enters into nerves and stays there for long periods of time. It does not replicate there, so T cells cannot recognise any antigen from the virus. If the person has a temporary lowering of immune resistance, perhaps through acute stress and/or sleep deprivation, the virus is reactivated and causes visible disease. Another member of the *Herpes* virus family, *Varicella*, is similar. *Varicella* causes chicken-pox, and after the first infection it remains in the body without causing any more chicken-pox spots. However, the virus can be activated when the new outbreak develops into shingles (Chapter 5).

It was thought that maybe the immunodeficiency viruses that cause AIDS in Man and simian immunodeficiency in monkeys also existed in a state of latency, up to many years in some instances, before any damage occurred. This 'silent period' as it is called should not be confused with the term seronegative used by clinicians to describe the disease's progress (see below). It is now known that this silent period is not a period of latency, but a time when the body's entire immune system is working hard to eliminate the virus, and how long it lasts will depend on the individual's ability to mount an immune response.

The time between infection and confirmation of this is also a period of apparent inaction by the virus, but again this is not so. If blood is taken immediately after a person is infected with HIV, and the serum tested for an immune response, it is generally found to be seronegative. This means that no antibodies against HIV can be found in the patient's blood. All the time, efforts are being made to improve the early diagnosis of infection by HIV, and the latest tests are sensitive enough to detect a change as early as two to three weeks after infection. Over the next two to three months, or even up to six months in some cases, the immune system responds and produces immunoglobulins so the patient is said to seroconvert from seronegative to seropositive. The patient may only have a transient non-specific glandular-fever-like reaction during this initial period, but there is usually some form of illness. After this, there is then a long period of time – the so-called silent period. This is when the body seems to win as no disease appears. As the virus gains hold, it begins to destroy cells with CD4 on their surface. These are primarily T helper cells and macrophages so the immune responses are weakened and other infections take hold quickly. Most of these are diseases that healthy people normally manage to suppress. These are called opportunistic infections because they can gain a hold only when a chink in the protective armour is found. With millions of people affected world-wide by the virus and the experience gained, it is now possible to design management therapies to suppress the virus, as well as to build up the immune system as a protection against opportunistic infections. As we will see in Chapter 9, such therapies at least in the earlier stages of AIDS generally include good diet with the right vitamins and micronutrients and a regime to maintain physical fitness.

With the more complex structure and life cycle of bacterial pathogens and more complex parasites, there are more opportunities for antigenic recognition by the host, and the activation of the cell-mediated immune responses. However, the more diverse the life cycle, as in many parasites, the more the parasites can vary their antigens to avoid detection. This makes it difficult for the individual to effectively eliminate the parasites and for society to raise good vaccines. Such problems exist with sleeping sickness which is caused by *Trypanosoma gambiense* and *T. rhodesiense*. This disease, which is transmitted by *Glossina* tsetse flies, affects wild animals and people living in parts of Africa. It is fatal to Man unless the patient is given drug treatment, and at present there is no vaccine. The difficulties in developing effective vaccines arise from the fact that the parasites have a surface coat which is made under the influence of one of about 1000 separate genes. The coat can be varied in an individual parasite by switching different genes on and off, so that new coat components are made. Thus the parasite is able to keep one step ahead of any antibodies produced by the infected person, and avoid being killed. This ability to modify surface antigens is a feature of many parasites and probably accounts for the success rate of these organisms. Some animals, for example antelopes in Africa, that have evolved side-by-side with the parasites are not affected. Man, who evolved later, and the modern domestic cattle brought into the area are often killed by the disease.

Even if the parasites do not change their antigenicity, some exploit the normal antigen/antibody response in other ways in their struggle to survive. In some stages in their life cycle, the malarial parasites *Plasmodium falciparum* have an antigen called S on their surface. If this is recognised by an immunoglobulin, the parasite sheds the S antigen before it is affected, and thus escapes damage from antigen-antibody reactions. This form of protection is simpler than the antigenic variations of the *Trypanosoma* parasites described above.

Tolerance is a lack of response by the host, not by the pathogen. Tolerance is needed by the body so that the immune system is not activated against self. The suppression of tolerance is necessary to allow organ and tissue transplants, and for the embryo to be accepted. It is thought that tolerance to self-antigens is achieved during early development. In addition the normal immune system can be made tolerant to a non-self antigen if it is given in very large doses, or if the baby meets the non-self antigen in the uterus. It has been shown that non-identical twin animals can become tolerant to each other's antigens if the blood in the placenta goes to both animals.

Both T and B cells can be made tolerant to antigens. It was originally thought that all potentially self-reacting lymphocytes were eliminated during development, but it is now known that they may exist in the normal adult in low numbers. It seems that other cells may keep them in check – T suppressor cells might be produced to switch off any auto-response, or additional signals needed for the self-reacting lymphocytes to be activated may be missing, perhaps because the body may have deleted the necessary T helper cells. However, it has been suggested that sometimes the tolerance that exists can be broken. This might occur if another antigen is very similar, and T cell help is given for that antigen. By chance this might also allow a reaction to be set off against self. This has been demonstrated after the use of certain drugs. For example α-methyl DOPA which is used to control severe high blood pressure can occasionally, in some people, cause an autoimmune-induced lack of erythrocytes. It appears that the drug may combine with the body's erythrocytes which then bear modified antigen. B cells of the body might then recognise this new antigen as foreign, and react to eliminate it from the body.

Although autoimmune diseases are probably caused in many ways, it can be appreciated that if drugs can create an autoimmunity, so perhaps can pathogens. In this case it is probable that the pathogens do not persist in the body, but that when they were present they caused an alteration in the immune system that persists long after the infection is over. If this is true, then it is not surprising that it is extremely difficult to prove that pathogens can be the root cause of some autoimmunities.

Although there are no definite links with viral infections in multiple sclerosis sufferers, it is one autoimmune disease for which progress has been made in identifying a virus as a contributory factor, and it also shows how complex the processes are. It seems that one virus in particular, adenovirus type 2, has stretches of amino acids on its surface that look just like a protein of the fatty myelin coat of nerves called myelin basic protein. In a normal immune response to a virus, antigen-presenting cells will take up the virus and process it to create fragments of viral antigen. These can then be placed on the antigen-presenting cell's surface membrane for presentation to T cells. The infected person's HLA-DR or DQ antigen receptor on T cells can see this viral antigen and will bind it. This will start the process of destroying the cell with the same antigen. However, T cells have now been activated against a short stretch of viral protein which looks just like myelin basic protein of the body. If the T cells find the body's own myelin basic protein, binding will activate the T cell and cause it to produce factors that attack the nerve tissues where the myelin basic protein is naturally present. It will also incite macrophages to perform similar cytotoxic actions. In addition, most studies have also found that the patient has an unusually high level of antibodies able to react with myelin basic protein – or autoantibodies. The complex of antigen and B cell autoantibodies can spark off complement-type reactions as described in Chapter 5. The end point is that a strong attack is made against the nerve sheaths, causing their death. Of course to do this in the brain the T cells have to get in. Normally only a few T cells patrol the brain, but in multiple sclerosis it is possible that cellular adhesion mechanisms are altered so the barrier can be penetrated and damage results.

Thus, even when the immune system appears to have overcome a pathogen, and eliminated it, there may still be remnants of that fight left over in terms of alterations to cells that can influence not only the speed of a reinfection, but also the body's responses to other antigens.

8

Survival of the Fittest

Alopecia areata It is said that the hair of Marie Antoinette turned white overnight before her execution in the French Revolution. How could this happen? Similar, but usually slower, changes occur in people who suffer alopecia areata or hair loss diseases. These are different from the hair loss described as male pattern baldness which is a normal part of ageing in men. In many cases of alopecia areata, the onset of disease follows illness or extreme stress. The exact cause of hair loss is not known, although recently it has been demonstrated that autoantibodies are present in disease sufferers, suggesting that the body is for some reason reacting against its own hair follicles and causing them to thin at the roots to form what are called 'exclamation mark' hairs. These weakened hairs fall out, or can be brushed out easily. Since it is always the pigmented hairs that are attacked, the white hairs of the head are left. The patient appears to turn grey rapidly, but actually the greying hairs were there all the time, but masked by the coloured hairs. In addition any hairs that regrow may now be white instead of the original colour.

Alopecia can appear in both men and women, at any age and over any hair-bearing site. There are several forms of the condition. The most common and mild is patchy alopecia. The hair may fall out or be brushed out in small patches on the scalp, and regrow again later. Alopecia totalis occurs when all of the scalp hair is lost, and in alopecia universalis hair is lost from all over the body. In some cases, there are no apparent reasons for the sudden loss of hair, but in others alopecia may be associated with autoimmune diseases such as systemic lupus erythematosus and pernicious anaemia. The condition can be helped with drugs, and spontaneous regrowth can also occur.

When most normally healthy children or adults contract an infection, the course of the disease and its recovery generally follows a predictable pattern. This is not so once the patient is less fit, as occurs in the elderly or those whose immune system is compromised by other factors such as drug abuse. One infection can follow another, and other parts of the body may be damaged by trying to fight the diseases. Fitness to resist infections requires a certain regime in life – regular healthy food and exercise, as well as the right life-style including a good attitude to challenges, but it also requires the luck of having healthy parents and, of course, enough money to avoid stress from financial and other burdens is a great help! How do these components influence good health?

A Sound Constitution

'A man is what he eats' is a famous quote (Feuerbach, 1850) with a profound meaning. To survive healthily, we must eat the nutrients that the body requires. Lack of the right food weakens the constitution and allows pathogens and diseases to get a hold. Whilst we all need a variety of similar foods, various cultures have met the body's requirements quite differently, so that it is quite difficult to lay out precise guidelines. When the whole variety of diet throughout the world is examined, it is obvious that there are a myriad ways of providing a good diet. The common components are proteins, such as meat and fish, carbohydrates like breads, cereals, sugars and cakes, fat in meats and dairy produce and fat used in cooking, minerals especially in fruits and nuts, and vitamins. In addition to being required for growth and successful reproduction, each of these food types has been shown to influence the immune system of the body.

One of the major problems of the world is protein malnutrition. With that comes an imbalance of the correct minerals, vitamins and fats. At the 1990 World Summit for Children at the United Nations, it was estimated that 40,000 deaths occur each day world-wide in children under five years of age. Whatever was listed as the cause of death, malnutrition was the common denominator. Disease is the usual killer in malnourished peoples, but not always so. A detailed study by 28 doctors working during the long Warsaw ghetto siege in World War II documented the effects of starvation on the siege victims. Some 85% of deaths came from infections in these people, so about 15% of those who died of starvation were disease free. However, in the absence of malnutrition, the body can mount a better defence against most diseases.

Very large numbers of children in Africa, parts of South America and the Indian subcontinent suffer acute malnutrition, Marasmus, or the longer term effects of an inadequate food supply which is called Kwashiorkor. The food, when available, is so poor that all the basic components of the diet are reduced, and some vitamins and micronutrients such as zinc may be completely absent. During pregnancy, the baby will take the nutrients it requires to the detriment of the mother's health. Once the baby is born, if the mother is starving and cannot supply milk then any baby of three to six months is especially at risk of being malnourished. When adverse conditions prevail, such as a prolonged drought, at least a quarter of all children are affected. There is now ample evidence that such people in these conditions also have poor immunity, especially cell-mediated or non-specific cell killing. They produce less IgA immunoglobulin, so their gut, lungs and other moist epithelial surfaces are not protected from infections in the normal manner. Resistance to pathogens is a major problem. Even when these children are vaccinated, they do not respond properly and do not gain any vaccination protection against the disease in question. This is particularly serious for measles since it has severe medical complications in such children. Another effect of poor vaccination success is that the measles virus can spread more rapidly throughout the population, and can infect babies too young to have been vaccinated.

T cells play an important role in cell-mediated killing, and the thymus, the key organ in producing these cells, is greatly reduced in size or atrophied in malnourished children. Thymic atrophy too is obvious in older people who die from malnutrition, in all societies. The thymus of children, especially the very young, is an enormously prolific producer of cells, and for this there is a large requirement for nutrients such as proteins, zinc and calcium. The lack of T cells from the thymus also means that the T cell regions of the lymph nodes and spleen are not well populated and it is therefore more difficult for an effective B cell immune response to be put into action. Without the basic building materials, cells cannot be made and disease cannot be fought. Since the basic problem for those with Marasmus and Kwashiorkor is lack of the right kind of food, access to a balanced diet can overcome the major problems of these illnesses.

Most information about the impact of protein in the diet on immunity comes from studies of animals given protein- and/or calorie-restricted diets and, in humans, from studies of malnutrition, voluntary fasting and the eating disorder called anorexia nervosa. In most cases where levels of protein in the diet are low, there is generally a reduced B cell-mediated antibody response whereas the more primitive and less specific cell killing or cell-mediated immunity is not always so affected. Even if total T cell numbers are reduced, cell killing can still be carried out by phagocytic macrophages or granulocytes. Fascinatingly, there is ample evidence to show that short-term fasting and controlled low-calorie intakes where there is no malnutrition or other dietary imbalances actually increase immune responses to a wide variety of challenges, including cancers, and result in a longer life. Clearly then, the key to good health is to keep a balanced diet without too high a calorie intake. Indeed in an unusual study of human starvation, when healthy young men had 50% less food than similar age-matched men, they reduced their body weight by 25% over a six-month period and were no more susceptible to disease than their heavier mates!

Young women are the commonest sufferers of anorexia nervosa. It is said that they perceive themselves as being overweight although to others they may not be. They therefore avoid eating anything that would result in weight gain. When this condition is mild, it is interesting that somehow, without doing it consciously, anorexic people do manage to maintain just-adequate protein, fat and vitamin intakes. Under such conditions, the body's ability to control infections does not seem to be greatly altered, and may even be improved. The full anorexic condition is characterised by changes in many body hormones, and unusual behaviours such as voluntary vomiting (bulimia) and abuse of laxatives, so the balance of food absorbed by the body becomes abnormal. Then, when weight drops below about 60% of normal for age and height, problems arise in fighting infections, and the body cannot mount good immune responses. Studies of wild bird populations have shown that death can be rapid when body weight falls below 60% of normal during quite short periods of adverse conditions that prevent feeding.

In many western countries and some societies where food is equated to wealth, overweight or obesity is a major problem. One-quarter of all Americans are severely overweight and the figure rises to 50% in some ethnic groups. Much attention is given to the bad effects of being overweight on heart function and disease, but such people also have an increased risk of diabetes, hypertension, cancers and early death. In addition, overeating and obesity can adversely increase the chances of getting an infection, lower the body's responses and increase the severity of illnesses.

What is a good diet? Most meals contain carbohydrates and fats in addition to proteins and bulk in the form of fibre. Those in the United States of America and westernised societies get about 15% of their calories from proteins, but in poorer societies carbohydrates may provide most of the energy. However, even when the diet contains a high carbohydrate content, certain components may be missing and render the diet inadequate. This occurs in Africa where the staple diet is often corn meal. This is deficient in the essential amino acid tryptophan, which the body cannot make and must take in from food. As a result, many people have Kwashiorkor. Tryptophan is one essential food, but some fats, vitamins and minerals are also essential. The unsaturated fatty acids such as linoleic acid, which comes from plants, is necessary in the diet to make a group of hormone-like substances called prostaglandins. These are needed for the correct functioning of many body systems. Other forms of fats, called saturated fatty acids, are present in animal fats. We need a mixture of both, but too high a saturated fat content results in a greater risk of getting heart disease. Therefore many people in westernised countries have moved towards eating purely unsaturated fatty acids. In America the trend is even more drastic, as many people cut out all fats of any description, and even go to the absurd lengths of giving their babies low-fat milk. This is not advisable for correct brain and good immune function. Fat is required to build and maintain the body. It is essential for the well being of the nervous system and for hormones and growth factors to be made. If monkeys are given a low-fat, low-cholesterol diet for over two years, the level in their brains of a substance called serotonin alters and they become aggressive. Low serotonin levels in Man have been linked to suicidal behaviour and low-fat diets favour depression. Diets of only poly-unsaturated fatty acids are inversely related to the useful activity of several immune system factors such as interleukin-2 and natural killer cells. With these extreme diets young women stop having periods, and babies fail to gain weight to the extent that they are described as growth retarded. Toddlers deprived of foods like eggs and meats do not get essential minerals and vitamins, and lack energy. Growing children need a high-calorie diet, balanced in all foods. Thus mixed diets with plenty of fresh vegetables, some fish or meat protein and some carbohydrate appear best for good health. The fresh vegetables and fruit supply vitamins, minerals and the bulk which gives good gut function. Too much fibre is harmful since high levels of dietary fibre mop up vitamins and minerals before they are absorbed, and result in low intake levels. It is the continued use of extreme diets that may lead to deficiencies of essential proteins or fats, or the existence of such a low carbohydrate content that body building proteins and fats are used up as a source of energy. That situation is dangerous as weight loss cannot be controlled.

In some parts of the world, the major foods available may present some risk of disease. Societies with a high intake of salted, smoked or pickled foods do appear to be at greater risk of developing stomach cancers, supposedly because these substance release nitrosamines in the stomach, and these are known to be carcinogenic (see Chapter 9). Other studies have led to the hypothesis that a high vitamin C intake via fresh fruit and vegetables, for example, might be able to inhibit the stomach from converting 'risky' foods into nitrosamines. However, whenever carefully defined populations of people are studied, there are other risk factors, such as the genetics of the group and environmental hazards, that could contribute to the increased levels of cancer observed. So the findings in one society cannot always be applied to another group of people in another country. Indeed, it is well known that emigrants from a population with certain low levels of susceptibility to a specific disease can, when moving to a new country, suffer the same or greater risk levels associated with that country. A well-documented example of this is a study published in 1965 of Japanese men who moved from their own country to the United States of America. Cancer of the stomach is much commoner in Japan than is cancer of the intestine, breast and prostate. Within a generation or two, the death rate from these cancers had altered, such that the number of deaths attributed to prostate cancer had increased while those caused by stomach cancer had decreased.

The soil we live on, and grow our crops in, is made of minerals. Extra high or low levels of certain minerals also influence how people cope with diseases. Some parts of the Middle East have naturally low levels of zinc in the soil and hence in many locally grown foods. This led to the identification of a disease called acrodermatitis enteropathica. Such zinc-deficient people, in addition to having many other conditions such as dwarfism and impotence, can be severely immunodeficient and cannot cope with infections. Clinically this is seen by reduced numbers and activity of T lymphocytes and neutrophils of the blood, or fewer cells to mount a good immune response. As in protein malnutrition, the thymus is greatly reduced in size, and so the spleen and lymph nodes have low numbers of T cells in them and macrophages do not phagocytose very efficiently. In addition, some immune factors and cytokines depend on zinc for their activity, and their reduced activity will also contribute to the poor immune responses. It has been noted for example that when the levels of zinc fall in people, so too do the levels of vitamin E. Chronic alcoholism, diabetes mellitus, gut malabsorption syndromes, very high fibre diets, certain drugs and many laxatives can all cause low zinc levels. From work with animals, it was possible to establish that it is the cell-mediated arm of the immune system that deals with cell killing that is compromised, whilst the B lymphocyte antibody responses are less affected. Putting zinc back into the diet fully restores immune function although the levels must be carefully controlled. Too little restores body weight and growth but leaves the immune system working under par. Very high levels suppress good immunity.

Another mineral deficiency, that of copper, carries with it abnormalities of the hair that make the disease easily identifiable. Copper deficiency though is quite rare. Much more widespread in the world is iodine deficiency. The thyroid gland needs iodine to produce the hormones that regulate our metabolism. The same hormones supply iodine to neutrophils to form a bactericidal compound. Thus people with abnormally low thyroid activity and those having iodine-deficient diets are prone to bacterial infections.

There is an increasing awareness today in westernised societies of the fact that many foods are prepared in advance, packaged, stored, transported and consumed much later than if they had come straight from the primary producer. This is generally regarded as beneficial to our societies, but is it? To what extent are we losing vitamins and micronutrients? Would optimally sun-ripened fruit be better for us? Do our diets need vitamin supplementation? Again, animal studies have furnished much information with regard to vitamins and immunity where deficiencies have been induced. Vitamins, by definition, are essential to life. Deficiencies therefore cause identifiable symptoms that are relieved by the addition of vitamins to the diet. We do not manufacture some vitamins in our bodies, so they must be taken in as food but levels of some vitamins, like vitamin C, are only high in fresh, not stored, food. Fortunately, some of the fat-soluble ones, for example vitamins A, D, E and K, can be retained in the body for months, but lack of the water-soluble vitamins, like the B complex and vitamin C, can cause deficiencies very quickly, even in days. Shortly after vitamins were named in 1911, the effect of deficiencies on death and resistance to disease were well documented. Since then, their importance for immune responsiveness has been studied, especially for vitamin A and its precursor form β-carotene, vitamin C, vitamin E, and the micronutrient selenium that can replace vitamin E in the body. However, it is only comparatively recently that the public have taken it on themselves to use vitamins as dietary supplements, although vitamin C in rose hip syrup and vitamins A, D and E in cod liver oil were given free to British children during and after World War II to keep them healthy whilst their diets were restricted.

Vitamin A is described as the anti-infection vitamin, because lack of it damages the epithelial cells of the body, so that infections quickly take hold and in extreme cases death results. In the 1930s in the United Sates of America it was established that the levels of vitamin A in children were directly related to their susceptibility to respiratory diseases. This was immediately followed by a trial in a large children's hospital in London, where measles-infected children were, or were not, given additional cod liver oil. A 50% reduction in mortality amongst those children given vitamin A was ascribed to the supplementation regime. Low levels of vitamin A occur world-wide and are often correlated with the state of 'night-blindness' or xerophthalmia. Because five to ten million children each year develop xerophthalmia, and a quarter of these become blind, most work has been done in this area. Many large-scale vitamin A supplementation studies have confirmed that vitamin A is used up very quickly when the body is infected, and that this allows survivors to be infected more easily with new pathogens. The pathogens are opportunists. Despite the advantages of vitamin A supplementation, it should be taken cautiously as it is toxic in large doses. By contrast, vitamin C, taken as tablets or a drink in abnormally high levels, seems to enhance the immune competence of phagocytes and to shorten the time taken to cope with viral diseases such as the common cold.

Other vitamins, especially vitamin E and the mineral selenium that acts in a similar way, are powerful agents for mopping up any free oxygen. This means that they can destroy harmful metabolic by-products in the body that cause cell damage. Oxygen in the air is in a stable form of two molecules linked together, but inside the body the two molecules can be split and free single oxygen is dangerous to cells. If the harmful chemicals are left in the body they can lead to the development of cancers. Large areas of the world are naturally deficient in selenium and their inhabitants may have related diseases; for example, the Californian seaboard in the United States of America, parts of western Australia, New Zealand and large regions of China where the deficiency syndrome is called Keshan disease. Keshan disease sufferers have more heart conditions, infections and cancers than their fellow Chinese living in places with normal selenium levels. In large parts of China where selenium is lacking it is given in salt to the entire population, and the incidence of disease has been dramatically reduced. However, high levels of selenium in the diet are dangerous to the body, and its use in toiletries such as shampoos designed to combat dandruff has been largely discontinued.

One of the precursors of vitamin A, a substance called β-carotene, is a strong antioxidant and has been associated with protecting children against respiratory diseases. It enhances the efficiency of vitamin E supplements too in giving resistance to disease. Abnormally low levels of β-carotene have been linked to cancers, but this needs further investigation. Fascinatingly, many anorexic children will make carrots, which get their colour from β-carotene, a staple part of their diet. They seem to instinctively know what they need to eat. This is rather like pregnant women who crave odd foods. Oysters and shell fish are rich in most minerals, and even the desire to eat coal has been explained by the body's subconscious need for iron and other minerals for the developing foetus. Acrodermatitis enteropathica patients living in zinc-deficient soil areas are described as often having an unusual habit of trying to eat soil. Undoubtedly our needs for different foods change when our body lacks certain substances. It is very difficult to prove that these food 'fads' are really linked to bodily needs. However, many people from cold or moderate climates who move to live in very hot countries know that they can take, and indeed want, very salty drinks that would probably make them feel sick under other circumstances. It seems as though the body can subconsciously direct our dietary intake, perhaps by subtle modifications of taste buds. Even the desire for eating chocolate has been accounted for this way!

The immune system is said to be a good barometer of the level of micro-nutrients in the body. Just as deficiencies contribute to illness and poor health, so too can excesses and imbalances. This is particularly true for some vitamins and minerals. Sometimes well-meaning dietary supplementation has led to increased susceptibility to disease. This happened when multiple vitamins and riboflavin were given to people in a region where malaria was endemic. The parasite burdens actually increased in some people. Apparently, a lack of riboflavin normally reduces parasite burdens. Good came of this however, because it drew attention to the fact that a deliberate reduction of riboflavin, sometimes caused by drugs, could be used against malaria.

The recommended daily allowances, or RDAs, for vitamins have been established over the years, almost by trial and error, and by assessing the credibility of reports. Exceeding the limits can, in some cases, be relatively harmless, but in others the levels are critical. Adverse effects are seen at ten times the RDA for vitamin A, whereas the levels of vitamin B6 may have to be 100 times greater for the same degree of toxicity. Vitamins E and C are considered to have intermediate values in this respect. However, since nearly all dietary minerals and most vitamins interact, the setting of safe limits is hazardous and dietary supplementation should be used cautiously.

A Fit Body

Is exercise good for the immune system, and/or longevity? There is no consensus of opinion. There is no doubt that people who exercise regularly tend to have a healthy life style, and many report a sense of well being and euphoria – the 'high' of runners. Regular exercise makes people feel well. However, the actual assessment of health in terms of incidence of acute illness and longevity is very difficult to achieve since there are so many confounding factors that ought to be studied at the same time. There is the type of exercise, its duration, the age, diet and fitness of the participants to consider, in addition to choosing how to measure immune function. It is well established that exercise damages muscles by the build-up of harmful chemicals such as some forms of oxygen. Normally there are antioxidants to mop up the free radicals in the blood. Vitamins E and A are particularly good in this respect. However, does this influence the immune system? All that is known is that some very well trained, or maybe overtrained, athletes have experienced fatigue, injury and an increased number of infections.

As will be seen in the next section, exercise is to be regarded as a stressor. That is, it causes the body to operate in a mode that achieves greater muscle power and cardiac output. These 'stress-induced' changes alter the numbers of immune cells in the blood and their relative proportions. This occurs relatively transiently, and could have completely different effects on the immune system to chronic emotional stress. Essentially both acute emotional and physical stress increase the levels of eosinophils in the blood. The effect of different types of stress was shown beautifully in a study performed on oarsmen, coxswains and coaches during training before and then after a four-mile Harvard-Yale boat race. A rise in blood eosinophils was seen in oarsmen but not coaches or coxswains during training, whereas it occurred in all three groups after the race. This rise, and other observations showing similar mobilisation of lymphocytes into the blood after exercise, has been attributed to the hormones of the stress response (see later). Thus, in general, in acute and short-term strenuous exercise undertaken by normally healthy young people there are shunts of mainly eosinophilic leukocytes and erythrocytes from tissues into the blood within 15 minutes. Whilst eosinophils are especially valuable in combating parasitic infections, their levels in the blood also rise in allergic responses. They may therefore be influencing the tone of blood vessels during exercise in a way that they would in allergic responses. This might also partially account for the increased numbers of erythrocytes in the blood, and would allow more oxygen to be taken to the muscles in a given period of time. With direct regard to immune capability though, exercise also mobilises more natural killer than B cells into the blood. As a result the numbers of some T lymphocytes decline so that there are there are fewer CD4 cells than CD8 cells. It has been suggested that the reduced numbers, and to some extent reduced activity, of these cells would tend to keep the total immune response capability at much the same level as in the non-exercising state. Generally, after the exercise stops the status quo in terms of the absolute number and proportion of leukocytes and their functional abilities is regained after about four to six hours, and nearly always within 24 hours.

The influence of exercise on cell mobilisation is seen more clearly as the level of exercise increases. Prolonging exercise beyond 30 minutes or so often results in another rise in blood leukocyte numbers. As the number of natural killer cells increase in the blood, so too does their functional ability. This was demonstrated in one group of elderly ladies with an average age of 72 years who undertook physical training for 16 weeks. Since the natural killer cells are important for immune surveillance, this increased function could be beneficial against cancers and infections. However, opinions differ as to whether the function of many leukocytes is diminished or increased with exercise, although it does appear that exhausting exercise adversely affects immune cells.

It has been repeatedly noted that there is an increase in zinc levels in the blood after strenuous exercise, and this has been attributed not only to shifts in blood volume and cells, but probably to a release of zinc from damaged muscles. However, it is complicated by the findings that erythrocytes actually lose zinc before returning to the pre-exercise condition. As we have seen, zinc is necessary for many cellular functions and generally enhances immune responses.

Athletes commonly suffer viral infections of the nose and throat. Indeed this causes more cancelled entries or 'below par' performance than all other diseases combined. The major disease-forming problem arises from contracting winter colds caused by rhinoviruses, although the common cold coronavirus of teenagers and the enteroviruses, especially coxsackie, are also damaging. Exercise during coxsackie infection can lead to heart damage, and it is reputed to be the cause of sudden death in strenuous exercise. It is easy to pass colds around in the crowded conditions of many changing rooms. Most athletes find that moderate training increases protection from colds, but there are clear indications that excessive training or overtraining, especially when coupled with mental stress, results in more infections. During a viral infection there are measurable losses in muscle efficiency, increased muscle aches, a raised heart rate, increased breathlessness and a feeling of weakness. The body seems to know best, because often the athlete does not want to train or participate. Some trainers use a 'neck check'.

Stuffiness and a runny nose above the neck indicate that work-outs should be done at half speed. If the symptoms disappear, normal exercise can be achieved but cessation is advised if the symptoms do not go away. Below the neck, problems such as muscle aches, coughing or fever mean that full training is only resumed after about 10–14 days. A good diet, no psychological stress and adequate sleep, as in all illnesses, hasten recovery.

Many athletes also, like many teenagers and young college students, may have glandular fever which is otherwise called infectious mononucleosis. This is caused by the Epstein-Barr virus which, like cold sores, is generally passed between people by kissing. About half of the people who get infected are not affected by it, but in the rest the infection holds back training and exercise for a long time, probably up to three months, before full recovery of performance. The incubation period is over one month long during which time the person may have headaches, fever, general malaise and even weight loss. Then, with the sore throat that develops, there is a great enlargement of the lymph nodes and by the second week an enlarged spleen. The virus attacks B lymphocytes and causes abnormal development. Fortunately the T cells are not affected and they set up an immune response resulting in the production of antibodies to the virus. However, this massive cell proliferation and cell death of the immune response cause enlarged lymph nodes and spleen. Usually the patient does not feel well enough to exercise, but if they do there is a risk of the spleen rupturing. No exercising for the first three weeks is the best rule. Athletes do not seem to show as many symptoms as non-athletes, but although the athlete recovers faster it still takes a long time to regain physical fitness after an Epstein-Barr virus infection.

The Stimulation of Stress

In discussing exercise, the concept of increased stress arises. Here the exercise creates a new challenge for the body – to provide enough energy to complete the movements. The body does this by switching to a mode that will mobilise energy quickly. This is the stress response or, in other words, how the body mounts physiological responses to demanding or threatening environmental influences. This is often described as ‘The fight or flight reaction’. We now know that mental stress and the stress of an immune response can elicit physiological changes in the body that are similar to those elicited by the more dramatic situation of being threatened by a man with a knife. In all cases of stress, early signals causing a change in the body’s physiology come from the brain. The autonomic nervous system, which acts without us really being aware of it, immediately makes nerves release catecholamines, namely adrenaline and noradrenaline (also called epinephrine and norepinephrine respectively). These rapidly make the heart beat faster, and blood pressure rises. The blood vessels to the muscles are relaxed in tone, so they carry more blood to the muscles. Thus the muscles have the energy supply required to cope with an intense burst of activity. Others blood vessels are reduced in size so that not so much blood reaches parts of the body such as the gut – no one needs to eat when running away! At the same time, but acting very slightly more slowly, chemicals released from nerve endings within the brain cause the hormone adrenocorticotrophin, or ACTH for short, to be released from another part of the brain. ACTH sends messages to the adrenal gland and, as a result, the adrenal glands release a flood of glucocorticoid hormones, including cortisol, and adrenaline into the blood. The glucocorticoids cause lymphocytes and granulocytes to be released into the blood from the tissues. So the heart pumps more blood around the body, the muscles have energy to respond and the whole body is primed to react to the stress. Thus fight or flight can be achieved. The speed of return to normal depends on the fitness of the individual, so training can allow the body to reduce and cope with the stress surge of hormones, and not suffer the sometimes dangerous consequences of stress. To a person with a weak or damaged heart, extreme physical exercise can be fatal. Stress can kill.

There are numerous forms of stress, and not all people are affected by a stressor to the same extent as others. In the 1970s a series of studies defined peoples' reactions to stress, and classified people as having type-A coronary-prone, or type-B non-coronary-prone behaviour patterns. Thus one's physiological make-up, age, sex, race, diet, physical condition and family history all affect how one responds to stress. All actors and public performers know it is important to feel 'keyed up' in order to give their best. Such mild stress is generally easily tolerated by the individual but, none the less, it evokes, to some extent, the stress reactions described above. Stressors are said to disturb the homeostatic or stable state of the body, so too do pathogens, and an immune response can be regarded as a stressor, and antigen as a specific stress stimulus. Indeed, the reactions after immunisation have been described as having many similarities with the body's reaction to physical and psychological stressors.

How does stress interact with the immune system? The release of hormones involved in stress come from the brain, so the brain coordinates the body's responses. Once the brain is alerted to stress, messages go out in the blood and nerves to reach all organs, including the lymphoid organs and immune cells moving around the body. As we have seen, one effect is to alter blood flow. An increased flow will allow cells with antigen to reach lymph nodes quickly to set in motion the B cell responses. Additionally it allows cells and antibodies in lymphoid organs to leave quickly and be in the blood to fight an immune challenge. Secondly brain-derived messengers talk directly to immune cells and cells that make up the lymphoid organs. Adrenaline can bind to receptors on the surface of these cells. Stressors such as exercising on a bicycle or restraining animals causes adrenaline receptors to be shifted to the surface of the cells where they can take up the circulating adrenaline more quickly. Under some viral attacks, some lymphocytes actually make ACTH and other factors formed from its precursors, many of which activate nearby immune cells. ACTH also has a direct action on the thymus, causing the release of hormones that kill thymocytes and affect how thymocytes develop.

The hormone ACTH is secreted rapidly, but the very slightly slower rise in the level of glucocorticoids, such as cortisol, in the blood will also be recognised by receptors on immune cells. For example, lymphocytes in the blood, natural killer cells, the epithelial cells of the thymus, thymocytes, and cells of the macrophage series have receptors for some glucocorticoid hormones and the family of catecholamines to which adrenaline belongs. Glucocorticoid receptors are of two types. Both bind cortisol but with different affinities, or strengths. This creates feedback mechanisms, so the release of cortisol is regulated. Cortisol binding to one receptor type is used for reactions concerned with inflammation and immune suppression, whereas cortisol bound to the other is inactivated quickly so allowing the receptor to function to influence blood flow and fluid metabolism by its binding with other corticoids called the mineralocorticoids. Like the sex steroid hormones receptors, to which the corticoid receptors are related, the receptors are inside the cell not on the cell surface, as are the adrenaline receptors and many of the other receptors described in this book.

High levels of glucocorticoids have long been known to adversely affect lymphoid cells, and to be involved in immunosuppression and an increased susceptibility to disease, as well as changes in behaviour and metabolism such as those associated with alcoholism and anorexia nervosa. Many animal studies have shown that thymocytes developing in the thymus are especially sensitive to raised levels of glucocorticoids in the blood, and they are killed by the process of apoptosis after their receptors are occupied. Thus T cell production is affected by a stress reaction of the whole body. Human blood lymphocytes are not generally killed by glucocorticoids, but the hormones dramatically affect lymphoid cell functioning. They are described as anti-inflammatory and immunosuppressive because of their actions on the growth, differentiation, distribution and function of cells in the immune system. In particular they suppress the secretion of cytokines and other factors from many different immune system cells. They have a profound effect on cell-mediated immunity, but can also modulate immunoglobulin production. The multiple suppressive effects of glucocorticoids are exploited in medicine by using them to reduce immune responses and dampen down inflammation. This mimics the normal actions of non-stress levels of glucocorticoids by causing lymphocytes and phagocytes to be 'marginated' or manoeuvred out of the way into tissues such as the bone marrow. They also stop cells from getting into inflamed tissues by altering the adhesion molecules on their surface. This keeps the cells in the blood; thus inflammation, for example in a damaged joint, is greatly reduced. A change in glucocorticoid levels in blood acts as a feedback mechanism to control the natural rate of their release. It has been suggested that in autoimmune reactions in the body, there are not enough glucocorticoids for the proper control of the immune reaction. It is known, for example, that people with the multiorgan autoimmune condition of lupus erythematosus have abnormal reactions to glucocorticoids, suggesting an inherent fault.

Thus stress reactions will alter immune functions. As scientists started to really understand the molecules used in immune reactions, it became clear that important immune reactions could be studied outside the body in the laboratory. This enabled the production of antibody and the factors that stimulate lymphocytes to be dissected in detail. A growing concept emerged of an autonomous immune system with principles that could be understood without reference to the other systems of the body. A wealth of earlier experiments and knowledge integrating the susceptibility to disease, and recovery from it, were dismissed as unscientific and irrelevant. At that stage in immunology research, very little was known about how the brain communicates with the organs and cells of the immune system. When in the early 1980s it was suggested that lymphocytes, after an immune reaction, can produce factors that are normally found in the brain, immunologists were in general sceptical that this would have any relevance to immunology. How wrong they were! It is now clear that not only does the brain communicate with the immune system, but the immune system talks to the brain. Intertwined with these pathways are the common use of many endocrine factors that modulate or enhance both the brain's and the immune system's responses to infection, disease and recovery. This has led to a new scientific discipline called psychoneuroimmunology or the relationship between brain, behaviour and immunity. By applying internationally recognised scientific techniques, new discoveries, explanations of observations dating back to the nineteenth century, and a better understanding of oriental medicine where there are different medical practices, have now emerged. Of particular importance are the actions of the part of the nervous system that is not normally under our conscious control, of stress and of behavioural influences on disease and immunity.

The nervous system is linked to all lymphoid organs by nerves, and the substances released by nerves directly bind to many immune cells. Damage to the brain has been shown to alter immune responses, and by studying these changes, there is a better understanding of the link between left-handedness and an increased susceptibility to autoimmune diseases. This is not a one-way interaction. It is now known that when an immune response occurs, the brain's pattern of neurotransmitter release is altered, and many factors from immune cells act directly on the brain to induce fever, sleep and a change in eating and other behavioural patterns. The use of these factors, for example cytokines, in medical practice has resulted in marked changes in mood too.

Both animal and human observations show that the degree of resistance to disease can be affected by stress. In the short term and in young people, the effects are less serious than in the elderly or sick. Students taking examinations are under stress, and this affects the immune system. They have fewer CD4 T helper lymphocytes, a poorer reaction by many immune system cells and an increased susceptibility to diseases such as glandular fever and *Herpes* infections. Most students recover quickly after the examinations are over, and no long-term impairment in immunity is seen. In other cases, such as the now very well documented accounts of bereavement associated with depression, there is often a greater risk of death and illness in carers and relatives, especially in older men. Many spouses of elderly couples die within six months of their partners. Marital discord, acrimonious divorce, personal and environmental disasters all evoke a greater susceptibility to infections and the reactivation of latent viruses. A common finding is that the blood contains fewer natural killer cells and often lower numbers of both T and B cells. This results in a reduced response to antigens. Depression reduces the natural killer cell numbers too. Natural killer cells are especially important for controlling the start of cancers, and a loss of natural killer cells in cancer patients is often life threatening. Depressed people often show evidence of viral infections that are not seen in similar but non-depressed healthy people or hospitalised patients. It is difficult to evaluate this type of observation though, as such changes could also be due to greater viral activity influencing the brain and causing clinical depression.

Despite all of these observations and many more, it is almost impossible with our current state of knowledge to predict how one individual will be affected by stressful circumstances. The outcome depends on the type and length of stress, how 'stress resistant' the person was previously, what sort of pathogen the stressed person is exposed to or has in his/her body, the environment around the individual as well as their age, sex and nutritional state. What has been found is that many people can be counselled to reduce stress thereby helping them to cope with disease. Patients receiving cancer therapy can be advised about relaxation and helped to accept the stress of knowing that they will die sooner than expected. Life is then often made more acceptable and even prolonged. Such is the basis of many healing therapies, many of which may not substantially alter the course of the disease, but help the person to cope with it by healing the mind. Relaxation regimes of many societies reduce the dramatic and harmful rises in stress hormones that appear in older age and some diseases. Finally, some recent research has shown that laughter can reduce stress in people of all ages. Controlling stress is within the reach of most of us, but it does require actions to be taken.

Inherited Health

The genes on our chromosomes that govern the body's development and function have come from our natural parents. They were inherited from one or the other parent, with or without change. It is not surprising therefore that there is often a family history of certain diseases. This has been clearly shown in the British royal family where the inheritance of haemophilia can be traced from its first occurrence in Queen Victoria's children. Haemophiliacs lack a factor necessary for blood to clot, so they are prone to bleed into joints and muscles when they suffer minor injury. Women are gene carriers, but it is only those men who inherit the gene who lack the clotting factor and are thus affected. In the generation after Queen Victoria, two women carried the gene and one man was a haemophiliac. However in the next generation there were three sufferers and three carriers and, because of intermarriage, the next generation had seven haemophiliacs and none of the women carried the gene. The haemophilia-gene-free arm of the family has continued through five generations without any disease.

Many autoimmune diseases run in families, although true brothers and sisters may not develop the disease at all, or to the same extent. In affected families, some members will have the autoimmune disease, others may develop different autoimmune conditions, and yet other relatives may not have any overt disease, although evidence for the individual's susceptibility may be found. There is an autoimmune condition called pernicious anaemia in which the erythrocyte count is low due to lack of a factor in the stomach that is necessary for proper erythrocyte development. In such families where pernicious anaemia occurs, some members may have instead an autoimmune disease that affects the stomach. However, the examination of the blood of everyone in the family often shows that some family members who have no pernicious anaemia or gastritis do have indicators in their blood that they could develop a different autoimmune condition which affects the thyroid gland. At the time of the examination they may not show or suffer any disease at all.

With the appreciation that autoimmunity often occurs in families, the search started for the genetic factors involved. Of great importance in this respect are the MHC genes (see Chapter 1) which produce antigens on leukocytes and most cells of the body. Much of our understanding of the mechanisms of MHC gene action in health and disease stems from studies of the mouse, where it possible to undertake many studies that cannot be done in Man because rodents breed faster and genetic mutants can be made and studied over a short period of time. Although the MHC genes occur on different chromosomes in mouse and Man, the genes are arranged in similar gene groupings, and their functions are very similar. The mouse MHC gene groupings on the chromosome are called H-2K, -D, -L and -I, and in humans A, B, C and D (Fig. 8.1). Some minor histocompatibility genes are localised elsewhere. In Man, the A, B and C genes direct the making of the most important class I antigens that are present on all nucleated cells of the body, and D is involved with MHC class II antigens on selected cells of the immune system, especially antigen-presenting cells. It is the antigen-presenting cells and other cells that bear MHC class II antigen, such as T helper cells, that show antigen derived from pathogens to the immune system – in other words, the form of MHC gene in the antigen-presenting cells will determine the type and efficiency of the immune response. The

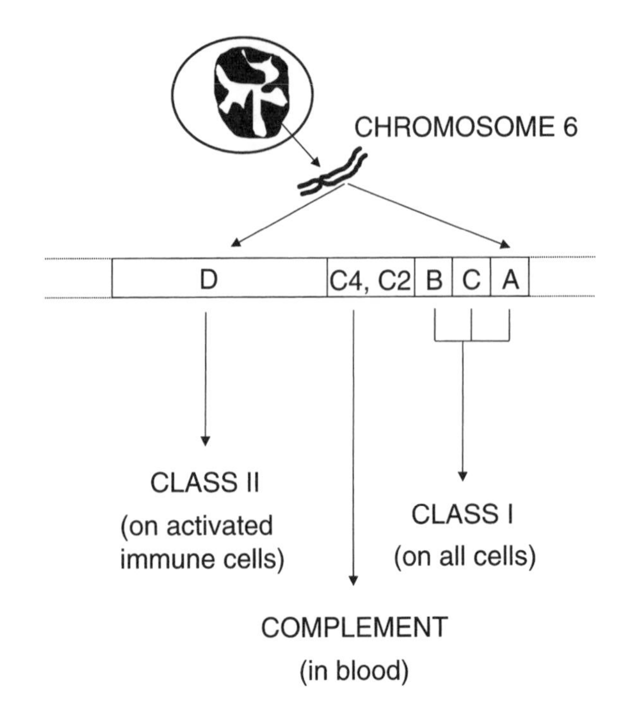

Figure 8.1. In an adult non-dividing cell, the chromosomes cannot be seen individually. They are part of the nucleus and, like other chromosomes, there are groups of genes along them that control the making of cell components. The A, B and C regions all have the information for making MHC class I antigens but the MHC class II antigen structure is controlled by the D region. Separate parts of the chromosome (C4 and C2) have the genes for making the molecules of the complement system.

MHC genes cause antigen to be cut, processed and represented on the cell surface in a specific way. People with different MHC genes will therefore present the same antigen in different ways.

It was soon realised that certain MHC genes, especially those of the D and B loci, were found more commonly in people with autoimmune conditions. The possession of a particular disease-linked gene does not mean that the individual will definitely get the disease, but there is a greater risk of developing it, compared with people who do not have those specific MHC genes. There are now about 20 autoimmune diseases that have been linked in this way, and some diseases show the association more clearly than others. For example, people with genes for HLA-DR4 are six times more likely to get rheumatoid arthritis than those without. Over 80% of people with the

HLA-B27 genes are some 88 times more likely to develop ankylosing spondylitis and also about 35 times more likely to get Reiter's disease than people without this gene. The genes for HLA-B8 or Dw3 are associated with coeliac disease, but individuals with the HLA-Dw3 gene could alternatively, or in addition, develop other autoimmune conditions, generally related to specific organs of the body, such as the thyroid-related Addison's or Grave's diseases. HLA-B8 genes are also found in sufferers of Grave's disease and active chronic hepatitis. It is not uncommon for one person to have several autoimmune diseases together.

Despite these strong indications of a genetic susceptibility, the development of an autoimmune disease must also depend on other factors. For example, identical twins with the same genetic background may not both get rheumatoid arthritis, which is generally associated with the DRw4 antigen. An identical twin of someone with multiple sclerosis has only a 30% chance of also being affected, but a 50% risk if the disease is juvenile diabetes. Even so, 5% of sufferers of this last disease do not carry the HLA-DR3 or DR4 genes for the disease-risk antigens, and 5% of the population carry genes for these types of antigen without having diabetes.

Finally, it would appear that, generally, adequate money means better living conditions, more attention being paid to health, eating and exercise, and sometimes less stress. In The Netherlands it was reported that people who did not go on to further education were three times more likely to be ill than those who did. The reasons behind this observation, if it could be substantiated in other groups, are complex and unlikely to be linked to just education. The whole life style is very important. It is said that money cannot buy good health, but it can help to keep it!

9

Struggle for Survival

Breast cancer This is the second most common cancer of women in the world, causing some 500,000 deaths each year. In North America, it carries a 12.5% 'risk rate', which means that about one in eight women will develop breast cancer, and each year some 46,000 will die from it. For the United Kingdom the rate is 1 in 12, or an 8% risk, with a similar proportion of women who die. Although detection rates have improved, and therefore the incidence appears to be rising, the death rates have remained stable. There are many factors that increase the risk of getting breast cancer, ranging from a woman's age, where she was born, her standard of socio-economic grouping, whether she is overweight, to family histories of breast and ovarian cancers. The fact that it often occurs around the age of 50, and is less common in younger women, may be related to the levels of the female sex hormones, particularly oestrogen, as they are growth stimulators. As breast cancer progresses so the cancer cells stop responding to oestrogens and become 'fixed' in a proliferating mode. The levels of these hormones wax and wane throughout life, especially through the menstrual cycle, pregnancy and the menopause, as well as with the degree of obesity. Women who have had their ovaries removed before the age of 35 years rarely get breast cancer. So oestrogen levels are sometimes manipulated to reduce the risk of developing breast cancer in susceptible women or those with some stages of cancer. However, no theories have yet answered the question of why women of Asia and Africa are less likely to develop breast cancer than those living in Northern Europe or developed countries. As people migrate to live in more developed countries so their risk of developing breast cancer increases for them and for their female children and grandchildren. Furthermore where there is a family history of breast cancer, especially in young women, there is an increased risk of a female relative developing it. In some of these cases, it has been found that the woman has inherited a susceptibility gene called *BRCA1*. This gene has the information for producing tumour-suppressor factors that regulate the growth of breast epithelial cells. Although about 1 in 300 women have inherited this gene, and sometimes a similar *BRCA2* gene, this does not mean they will get breast cancer. For breast cancer to develop, the gene must change or mutate. Such mutations have been found in about 7% of women with a family history of breast cancer. Once a *BRCA1* mutated gene is present however, there is an 80–90% risk of developing the disease. Originally it was thought that the incidence of mutations was up to 80% in women with both breast and ovarian cancer, but it was later realised that these figures referred to a small group of patients who do not represent all those with breast cancer. However, over the last few years, efforts have been made to screen people to see if they carry these genes. The genes cannot be altered or removed, but the woman can be carefully and regularly examined by a specialist so that any small lump developing in her breast can be removed early, and relatives alerted to the possibility of developing breast cancer too.

Early detection is very important for long-term survival. The smaller a breast cancer is when it is found, the less likely it is that it has spread into nearby cells and the greater the chance that the cancer can be controlled. A few women with mutations of the *BRCA* genes have even undergone removal of their breasts to prevent them getting any cancer, but this is a drastic step to take and it may only benefit a few women who might have contracted cancer when they were under 40 years of age. It is more common to use surgery, preferably carried out during the low-oestrogen period of the menstrual cycle, combined with drugs that reduce oestrogen levels in the young (these may induce an early menopause), or prevent the breakdown of other hormones to form oestrogen, in post-menopausal patients.

Cancers and their Incidence

What are cancers? Cancers are a family of diseases that can afflict any part of the body. They all share in common an abnormal pattern of growth that causes cancerous cells to dominate the normal functioning of tissues and organs, and to spread out to invade other regions of the body. They start from one cell, or a very few, in which the DNA has been damaged. If these abnormal cells can escape the normal control processes of the cell, then they proliferate to form growths or tumours. Normally cells with damaged DNA would be killed or their DNA repaired, so such cells should not survive in the body. Occasionally, some cells are not killed, but the growing cells are not cancerous and the cells do not behave or spread out like cancers. Such growths are called benign tumours. However, benign tumours may cause problems in the body, by constricting or pressing upon nearby normal tissues. When cancerous cells escape the normal controls of cell proliferation, they often divide faster than normal and generally do not differentiate into mature cell types. They develop different recognition markers on their surfaces, and do not make the same molecules or secretions as the cells in the organ they came from. As cancerous cells accumulate, they show an increased ability to spread away from the primary tumour. Once in the blood or lymphatics, they can be carried to any organ, and having invaded other tissues they form localised accumulations of cancerous cells called metastases. Normal cells, other than erythrocytes and immune system cells, do not behave in this way; they stay within their tissues. If cells of the lymphoid system and phagocytes become cancerous, this is often seen as an enlarged and malfunctioning lymph node, with the cancer later involving the liver and spleen too. Hodgkin's lymphoma is a common cancer in young adults, but it may affect people of all ages. This cancer may start in one or two lymph nodes, perhaps in the neck or chest, but spreads to the lymphoid organs, sometimes including the bone marrow. This cancer can be quite fast to develop into a serious life-threatening condition but, if identified early enough, treatment by radiotherapy and/or chemotherapy may cure two out of three sufferers.

It is generally accepted that the development of cancer is not the result of just one cellular change, but is multifactorial. The concept of an initiator and a promoter is used to explain how an initial early insult to a cell can, many years later, be triggered by another event to create a cancerous condition. The initial trigger can be due to factors associated with a family history of cancer, a viral infection resulting in DNA alterations, or chemical damage to DNA through habits such as smoking or occupational exposure to dangerous substances. Diets low in vegetables and high in fats have also been implicated in being causative, but these could be a result of a lack of protective factors rather than an instigator of DNA damage. Generally, after the initial insult, there is a long interval of some 15–30 years when the damaged cell(s) is present in the body, has escaped surveillance, and is quiescent. Perhaps, like the early stages of HIV infection, this is a time when the immune system is working hard to keep the malignant cells in check. We do not know. All we do know is that, many years later, something triggers cancer development so that the cancerous cells get the upper hand, and disease progresses. Many people believe that triggers are linked to periods of extreme stress such as divorce, bereavement or some series of events that lower the general immune responsiveness of the individual. This is difficult to prove conclusively since so many factors affect an individual's chances of getting cancer. In young and normally healthy people immune surveillance is acting all the time to seek out and destroy abnormal, infected or diseased cells, but as one gets older or is for some reason immunocompromised, as are those with AIDS, so the chances of a cancerous cell surviving in the body increase.

The patient can often be completely unaware of this dangerous change in activity. Cancers frequently grow silently – initially there are often no specific symptoms or pain, perhaps some loss of energy or weight loss.

Clinical examination at this stage will then usually reveal a lump or tumour at one site. At some juncture, the cells at this primary site manage to escape from it, spreading into surrounding tissues or the nearest lymph nodes. The cells may get into blood vessels and lymphatics, and then be carried to more distant sites and organs to form metastases. Some cancers spread rapidly around the body, and such aggressive cancers are the hardest to control. Others may remain relatively confined in one site, and it may be possible to surgically remove these so that the patient has a very good chance of totally recovering. Cancers are therefore classified according to which organ they originally came from, which cells are involved, if the tumour is benign or malignant and how large they are. From this examination, doctors can give a prognosis or forecast of the outcome after treatment. The smaller, more localised, slowly growing tumours without metastases can usually be surgically removed with a very good prognosis, which may include a complete cure. Nevertheless, the patient must realise that even when all of the visible cancer has been removed by the surgeon, unfortunately the cancer may recur sometime in the future because a few microscopic cells, which we cannot detect with present methods, have already moved away from the primary site. Prolonging survival time depends primarily on successful treatment, although it now appears that a minority of people are better at fighting cancers or denying their presence than others. In most cases, good counselling helps people to adjust to the diagnosis better, and may improve the quality of life for those with cancer and other serious diseases. Although there are documented cases of complete recovery without surgical intervention or chemotherapy, these are extremely rare. Most people need surgical and medical help to try to stop, or slow, the spread of the cancer.

Between 25 and 40% of all cancer deaths are caused by smoking. Some 80–90% of smoking-related cancers are lung cancers, but smoking-related cancers also occur in the mouth, pharynx, oesophagus, stomach, intestines, pancreas, bladder and cervix of the womb. Twice as many men than women die of lung cancer, but women are catching up fast as more young women smoke now than previously. Stopping smoking has an immediate effect, and lowers the risk of developing cancer. The longer one has stopped smoking, the lower the risk of getting cancer. Many women in the UK die of breast cancer, but its cause is far harder to pinpoint. Skin cancers are the next most common forms of cancer, both in men and women. The majority of skin cancers can be treated, but some 20% are due to malignant melanoma. This is harder to treat as it spreads out as metastases, early in its growth and it is resistant to conventional chemo- and radiotherapy. The next most common cancers are those of the prostate gland in men and bowel cancer both in men and women. All of these tend to be commoner in older people. Prostatic cancer seems to be on the increase, but fortunately most prostatic cancers are slow growing and patients can survive with this form of cancer into old age. More men die with prostatic cancer than of it.

The patterns of cancer incidence are not the same in all countries. For example, the major cancer killer amongst men in the United States of America is lung cancer, but prostatic cancer is the next. Breast cancer claims more lives than any other cancer amongst American women. In Japan lung cancer is not as prevalent as in many other countries but smoking is increasing in popularity and the numbers affected who will die of smoking-associated cancers are rising. Japan has more cases of stomach cancer per head of population than westernised countries so they have instigated an intensive screening programme to pick up stomach cancer in its early stages when it can be successfully treated by surgery. Quite often when it is diagnosed in a patient in the UK the cancer is too extensive for surgery. Aided too by the changes in dietary habits, the incidence of stomach cancer in Japan is now declining. The influence of diet on stomach cancer in migrant Japanese moving to the United States of America who adopted the dietary habits of the new country was alluded to in the last chapter. The effect of diet on the incidence of cancer was also observed in Africa. The high-fibre diets result in faecal matter being passed rapidly down through the body. This means that cells lining the gut are not exposed to any carcinogens in the diet for as long as in people who have low-fibre diets. Such high fibre eaters generally had a lower incidence of bowel cancer than westernised peoples eating more mixed and less fibrous diets.

Normal and Abnormal Cell Cycles

In order to understand how cells can become cancerous, it is necessary to describe in some detail the sequences of events that occur when the cells of tissues and organs divide. Most cells divide to produce two new cells each of which contains exactly the same number and type of chromosomes as the original cell. The exception to this form of division, which is called mitosis, occurs during the formation of eggs and sperm when the cells divide by another process termed reduction division or meiosis. A different form of cell division is necessary because each egg and sperm only needs one set of chromosomes from the parent. Then, when they unite, the newly fertilised egg will have a set of chromosomes from the mother and one from the father.

Generally cells are in the non-dividing state, and when division does occur it takes place very quickly. How often cells divide depends on whether the tissues are growing in the embryo or in the adult, whether large numbers of cells are needed by the body and whether individual organs are damaged and need repairing. In general, cell divisions occur more rapidly in the embryo. Indeed, in the adult some cell types such as nerves, skeletal muscle cells and erythrocytes do not divide at all when mature. Nerves once formed are there for life. If damaged, the ends of the cells, which form the long fine nerves that go to all parts of he body, can be repaired. But if the cell body of the nerve containing the nucleus is damaged, then the nerve dies and new nerve cell bodies are not made to replace the original one. Erythrocytes, on the other hand, only live for 120 days, so new cells must be made all the time to keep the numbers of erythrocytes constant. The thymus, in young people, produces millions of cells to form the T cell repertoire, but 95% die before they get out into the blood, because they do not have the right T cell receptor. The cell divisions in the thymus are rapid; occurring about every eight to ten hours. With increasing age, fewer mitoses take place, but the cell still takes the same time to divide although the intervals between divisions may be longer. The lining of the gut also need a large number of new cells all the time. In the small intestine, which is the primary site for food digestion, the cells are damaged and destroyed very quickly. Here, cell-cycle times range between 11 and 24 hours, and a cell's life span is only about three to five days. This rate of cell mitosis is maintained throughout life. In all cells though, as the chromosomes in the nucleus are replicated in the division process, so a small change occurs in each chromosome that signals how many times the cell has divided. There seems to be a finite number of divisions possible for many cells, and as yet no one has managed to extend this in Man to prolong life.

The cell cycle is divided into four successive phases which are shown in Fig. 9.1. The stage called mitosis, M phase, is when the chromosomes in the one cell are replicated and then shared out between two new cells. The next three phases in each of the newly formed cells complete the cycle, and together they are called the interphase period. Thus mitosis and the interphase period equal one cell cycle. Immediately after mitosis, the first of the three stages in interphase is called G_1. This is a time when the cell uses nutrients to grow and attain normal size. After this, the next period – which is called the S, for synthesis, phase – starts. During the S phase, the DNA in the nucleus is doubled so that when the cell undergoes mitosis later, each

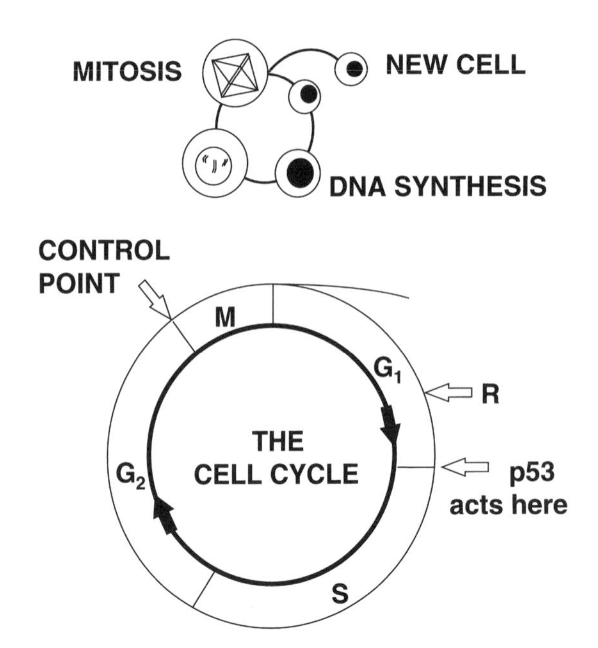

Figure 9.1. The cell cycle consists of a phase when deoxyribonucleic acid (DNA) is replicated (S) and a phase when it is halved (M) so that each new cell (released from the cycle at the start of G_0) has the same amount as the original cell. Restriction points (R) occur in the gap periods, and the guardian of the genome (p53) acts between G_1 and S phases.

new cell has the same amount and kind of DNA as the original cell. The third phase is called G_2 and is between the S phase and mitosis.

In general, when cells have a long cell-cycle time, they stay in G_1 for longer. The actual time taken for cells to go through mitosis, the M phase, is often around one hour, even in cells with a very long cell-cycle time. The G_1 phase, which can be of variable length in different cells, has a point towards the end known as the restriction point or R point. Cells that reach this point go no further and stop growing, although they still synthesise cellular components and function normally. Cells can stay healthy at this stage for a long time, even if starved. If cells are starved at any other point in the cell cycle, they die. If cells go beyond the restriction point they have to complete a full mitosis cycle and become two new cells. Thus theories have developed suggesting that this restriction point is actually a measure of the amount of a certain U or unstable protein produced in the cell. If the protein is unstable, unless a lot is made quickly, there is never enough to take the cell through the R point. Such a single protein is hypothetical, but a suppressor gene product, called the retinoblastoma gene product or Rb-1, does help to regulate cell-cycle progression at the critical G_1 restriction point. Rb-1 interacts with a system of proteins called cyclin-dependent kinases, which are implicated in allowing the progression of cells through the R point, and indeed through all four phases. There are lots of cyclins and kinase subunits, but their genes are only activated for short periods of time. Thus variations in the concentration of different cyclin kinases could control the cell cycle, and afford a means of 'feedback', or the ability to regulate the processes. However, as in all life processes, complexities arise. The cyclin genes are themselves controlled by a number of other factors, so what is the ultimate control? We do not know, but it is not likely to be a single substance, event or stage in the cell cycle. Multiple factors will all contribute to the control of cell cycles, and the large family of tumour suppressor genes are very important. In all cases the proteins produced under the control of these genes have a role in the normal regulation of cellular proliferation or differentiation. Tumours arise when these roles are perturbed and cell growth is not constrained by the regular processes.

The result of having a cell cycle is the creation of two new cells. To achieve this, it is necessary that DNA is replicated. This happens in the S phase. This replication must give two true copies, or the cell line would not be continued as a clone of the earlier cell. If a mistake occurs, the cell is said to have mutated or changed its chromosome composition. Cells have checks built into the cycle at specific points to stop mutants surviving. There are checks on DNA at the G_1 to S transition, and another at the G_2 to M change. If a cell, during its life, has damaged DNA, an increased level of a protein called p53 is made under the direction of the *p53* suppressor gene, and the cell is held dormant at this stage (in G_0), or can be triggered to undergo apoptosis and die before DNA synthesis can occur. In this way, no new replicate of this damaged cell is made. Similarly, genes are activated to produce other controlling proteins at the G_2 to M transition if the DNA is damaged or not properly replicated. However, much less is known about how this checkpoint works, whereas the p53 protein has been greatly studied in normal and cancerous cells. More will be said about the implication of p53 in cancers below.

After the DNA has been copied and the cell divides, the newly formed daughter cells are smaller than normal and there is therefore a period of growth. All cellular components must be made, and this occurs during G_1. All cells need extremely minute amounts of specific growth factors to grow, to divide and to survive. Studying growth factors has led to some of the most exciting breakthroughs in the study of cancer. During work on viruses that cause cancer in animals, it was realised that there is a group of genes that control a series of master switches during normal growth and development. These genes are called oncogenes or E genes. During a viral infection, the viruses switched on an oncogene, and the infected animals developed cancers. This led to the appreciation that the many different causes of cancer, by acting through oncogenes, act in a similar way at the DNA level. Activation of an oncogene stimulates the cell to release a protein whose function is to trigger growth and suppress apoptosis. Thus in viral infections the normal controlled cell cycle becomes uncoupled, so uncontrolled growth and proliferation result. Hence oncogenes are often referred to as cancer-causing genes. The *p53* and *rb-1* suppressor genes referred to above can suppress the action of oncogenes, and activate apoptosis.

There are many growth factors. Most eventually affect the nucleus, although they may bind to surface or cytoplasmic receptors. The sex steroid hormones, such as oestrogen and its derivatives, enter the cell across the cell membrane and bind to cytoplasmic receptors. Once bound, the complex moves to the nucleus and tacks onto the DNA in the nucleus inside the cell. Steroids can therefore switch genes on and off, and so change the way the cell will develop by directing which gene products are made. Many of the surface-acting signals are small peptides derived from proteins that act, by binding to their respective peptide receptors, to set in motion a cascade of second messenger events inside the cell initiating growth or cell division. These growth factors control a number of growth switches inside the cell. Such factors can derive from nearby cells, and are cell density dependent. When cells are well spaced out, especially in culture dishes in the laboratory, they seem to have a greater tendency to divide and grow, but as cells get crowded together they slow up and stop growing. Similar mechanisms could be used by organs to regulate their size, as normally there is very little change in organ size throughout life. The position of the cells relative to other cell types, and to connective tissue components also has an influence on how cells divide. For example, in the skin the only layer of cells to divide is that closest to the underlying connective tissue. Once the cells divide, they are pushed out towards the surface and away from the influence of the underlying connective tissue. With so many signals from different aspects of the microenvironment, it is not surprising that the system can get out of control and cells become cancerous. Indeed, as the processes controlling cell cycling are being unravelled, it appears that many carcinogenic processes are related to abnormalities in the normal patterns of cell growth and division.

Despite the precise sequences of events in the cell cycle, and many factors participating in controls to ensure that when cells divide they are faithfully copied, mistakes do occur. However, it probably requires more than just one mistake in cell division to create a cancerous cell, and a number of different types of intervention appear to be involved. Firstly the inherited genes may, in some cases, contain genes that have a greater tendency to mutate or change than others. Then there is the possibility that when changes occur in the genes the new situation may favour loss of, say, cell-cycle control, so allowing unregulated cell proliferation. This could occur when infectious agents such as viruses get into a cell and thereby change the genetic composition of that cell. Then there are a large number of environmental or occupational hazards that damage cells. In many cases the damage can be repaired, but if the cell-cycle control mechanisms are already altered by mutation or infection, then the double insult to the cell may be enough to allow it to become cancerous.

A most important gene in health and disease is that which controls the production of a protein called p53 referred to earlier in this chapter and in Chapter 7. This protein in the nucleus has been called the 'guardian of the genome' since its main function is to bind to and modulate the expression of genes controlling DNA repair, cell division and cell death by apoptosis. The *p53* gene is one of the tumour suppressor genes, so that damage to this gene not only results in loss of tumour suppression, but also misregulation of all the other genes to which p53 would normally bind. Cells with damaged DNA rapidly increase the levels of p53, and this causes a factor to be released that inhibits the cyclin-dependent kinases needed for cells to go through the cell cycle. Thus cells with damaged *p53* genes or without functional p53 proliferate more and cannot repair damaged DNA. At least half of all malignant tumours have mutations or rearrangements of the *p53* genes, and a rare group of people suffering the Li-Fraumeni syndrome have inherited mutations of the *p53* gene and develop many different cancers. Such tumours with defective *p53* genes are aggressive. They are fast growing, readily invade nearby tissues and are resistant to therapy. They are associated with 80% of colonic cancers, 50% of lung cancers and 40% of breast cancers.

Oncogenes, as we have seen above, are important controllers of the cell cycle. Certain viruses contain oncogenes and these can disrupt the cell cycle by various mechanisms. In order that viruses can reproduce, they insert their DNA into the host's chromosomes. This can cause nearby genes to have an altered function. This is called insertional mutagenesis, and the changes are inherited by the following generations. Another way in which viruses can cause cancers is because they themselves actually contain oncogenes. As we have seen above, oncogenes are central to controlling cell cycling, so it is easy to appreciate that extra oncogenes, especially if switched on at inappropriate times, could cause uncontrolled growth. It seems too that some oncogenes can prevent cells from undergoing development into their final form, or can cause cells to remain undifferentiated. Typically, cancer cells are undifferentiated. Thus oncogenes in viruses would ensure that the host's cells are kept alive, with the result that the viruses survive.

Whilst classic oncogenes accelerate the rate of cell proliferation, others are more involved with cell death. Some of these belong to the *bcl-2* family of genes which were found to be responsible for the development of B cell lymphoma/leukaemia-2. The actions of *bcl-2* genes were highlighted in studies of follicular non-Hodgkin's B cell lymphoma in which the *bcl-2* genes were changed so that the cancerous cells were dysregulated and produced too much Bcl-2 protein. High levels of this protein prevent cells from entering the apoptotic pathway of cell death. The gene c-*myc*, which encodes another oncogene product and accelerates cell proliferation, is also altered in some aggressive lymphomas and leukaemias. It is thought that the combination of preventing cell death in damaged cells and enhancing proliferation is one of the reasons why these cancers are so hard to control.

Carcinogens and Chemicals

Factors that can alter DNA are called carcinogens. Identifying carcinogens is a slow process that has to be done very thoroughly. It involves many tests including applying the substances, at many different concentrations, to bacteria or cells grown in culture, and then to animals. As we have seen, non-carcinogenic damage to DNA may occur to our own cells during cell divisions, and each cell contains the means of repairing such damage. When the damaged DNA cannot be repaired, then there is a potential for these cells to enter uncontrolled growth. Usually, the development of a cancer is considered to require irreparable injury to the cell, plus other factors that may stimulate the cell to behave in a dysregulated manner. This is called a multi-step process.

Carcinogens can come in many forms. It has been estimated that 70–80% of cancers are caused by factors in the environment, and some 10–15% by viruses. Ionising radiations generally induce leukaemias in a very short period of time. After the Hiroshima atomic bomb, there was a rise in both acute and chronic myeloid leukaemias of the bone marrow for about 14 years, with most cases occurring six to eight years after the explosion. Fortunately it seems as though after some ten years or so, the risk of radiation-induced leukaemia decreases. The length of time taken for the leukaemia to develop depends on the dose of radiation received, and on the timing of the exposure. The use of X-ray imaging to diagnose disease is now carefully monitored, and exposure is kept to a minimum. From past experience it was found that diagnostic radiographs used for examining babies in the womb increased the risk of childhood cancer by some 50–100%, and is therefore rarely undertaken nowadays. It seems that fast growing tissues are the most sensitive, and children up to about two years of age may develop cancers some two to ten times more often than adults receiving the same radiation exposure. In the past, people given X-ray treatment for a form of arthritis of the spinal column called ankylosing spondylitis died more often from blood cancers than untreated people. Not all radiation is experienced clinically as some radioactivity can comes from natural sources. Even in the sixteenth century, it was known that miners in some German and Czechoslovakian mines suffered a high rate of cancer and lung disease. Radioactivity from radium, which occurs naturally in granite, sandstone and limestone rocks, is regarded as a possible source of cancer, especially in babies and children. In the UK, for example, this hazard could affect people living in poorly ventilated houses throughout the southwest of England, hilly regions of the Midlands, Wales and Scotland. Another natural risk is ultraviolet (UV) light which causes breaks in DNA in a specific manner. The types of UV-induced damage can normally be mended by the cell, and results in a burst of DNA synthesis that is unusual in that it is unrelated to the process of cell division. Skin cancers occur mainly in fair-skinned individuals, and they are caused by long unprotected exposure to the sun. Burning by the sun is particularly dangerous. Modern sun screens have been developed to reduce this risk, but it is thought that with the depletion of the ozone layer that is now occurring in parts of the world, the risk of damage is increasing. Again children are particularly vulnerable, and need to be well protected from the effects of the sun. It is accepted that virtually all cancers of the skin of the face amongst white races are due to UV-induced damage. Malignant melanoma is another cancer linked to exposure to UV light although there may be other factors that contribute to its development.

The most well studied of all links between chemicals and cancer is the relationship between smoking and lung cancer which is of higher incidence in men than in women. The link was first shown in 1939, and since then numerous studies have been conducted on lung cancer. Nowadays the proportions are changing as many more women smoke today than in previous years. Thus, now, more women die from smoking-related cancers than breast cancer. There is a relationship between cancer and the number of cigarettes smoked daily, the length of time for which the person has smoked and the age at which smoking was started. A recent study has shown that a small number of cigarettes smoked over a long period is more harmful than smoking greater numbers over fewer years. Thus to reduce the risk of getting lung and other cancers, it is more important to stop smoking completely – there are no mitigating reasons for smoking only a few cigarettes daily! From several studies, it has emerged that lung cancer, bronchitis and emphysema, laryngeal cancer, cancer of the mouth and oesophagus, bladder cancer, coronary artery disease and renal cancer all caused death more frequently (in the above order of importance) amongst smokers than non-smokers. Indeed it has been suggested that in the United States of America, 40% of all male cancer deaths between the 1940s and 1970s were due to smoking. As usual, the situation is complicated because some of these cancers are commoner amongst alcoholics, who may also smoke. It does seem that both heavy smoking and heavy drinking greatly increase the risk of, for example, cancers of the mouth region.

Many other chemicals are carcinogenic, and occupational exposure can be a serious and long-standing problem. Generally, the skin, respiratory tract and the bladder are most commonly affected due to the manner in which the carcinogen enters or leaves the body. Cancer of the scrotum was first noted amongst chimney sweeps in 1775. Modern Health and Safety Standards in westernised societies are reducing these risks, but because of the long latency, and therefore problems of being sure which substances are carcinogenic, it will take a long time to eliminate some of these risks. Coal, soot and petrol products are high on the danger list, as is asbestos and mineral oil as used by machinists and spinners. This latter substance caused Shale-oil cancer on the hands and arms of Scottish workers. The substances in Coal tar that cause cancers are aromatic polycyclic hydrocarbons, which become carcinogenic when they are metabolised in the liver. Because of differences in tissue metabolism, diet and infection there is considerable variation between individuals in how they cope with these substances. The genetic make-up of some people makes them much more able to detoxify drugs and toxins than others and, as we have seen, if there are faults in the inherited genes that control cell cycling and/or cell death, then these individuals will have a greater risk of developing cancer following exposure to environmental or chemical carcinogens. Thus exposure to these potentially harmful substances may produce cancer in one individual, but not in another.

With the increasing use of motor vehicles, there is now an awareness of the irritant and damaging effects of petrol and diesel products. Standards of exhaust emission are now more controlled in many countries, but where there are strong cost disadvantages in adopting new standards, progress in controlling pollutants is often slow. Certain types of asbestos, especially blue, are carcinogenic and special precautions are now defined for demolishing old buildings where asbestos may have been used. The ways in which chemical carcinogens alter DNA are varied, but generally they either bind to the DNA and interfere with cell division or gene reading, or disrupt the DNA on a 'hit and run' basis. Fortunately, as we have seen above, the cell has an array of methods available for repairing damaged DNA or eliminating damaged cells.

Carcinogenic factors in the diet are now receiving more attention in medicine. There are a number of natural foods that contain carcinogens, but with the US Food and Drug Agency allowing some 2700 distinct chemical entities, chemical mixtures and plant extracts to be added to food, there is concern that consumption of some of those known to be harmful to health might increase the risk of developing cancer. Some of these, for example nitrites and nitrosamines, which have been firmly identified as carcinogenic, are now no longer used as food additives. Unfortunately the body has enzymes that change nitric oxide into the cancerous M-nitrosamines, so the risks from these substances cannot be eliminated. A positive correlation has been found between fat intake and cancers of the breast, uterus, pancreas and colon. A high salt intake has been linked to gastric cancers. However, studies aimed at analysing the diets of people living in areas of high cancer occurrence have failed to provide any strong evidence for any particular risk items in the diet. Some of these studies, for example those in the Linxian region of China and in Korea, have been very large and exhaustive. From other studies though, there are indications that diets predominantly of smoked meats or fish, especially when few vegetables and fruits are eaten, might increase the risk of gastric cancer. It is thought that the risk chemicals in smoked foods are nitrosamines, but recently other more potentially carcinogenic factors have been identified by heating meats and fish. Some of these are similar to carcinogenic factors in tobacco smoke, but many more studies need to be done to evaluate their potential risk to Man. It does appear from several studies that having plenty of fresh fruits, vegetables (especially some types of onions), antioxidant vitamins A and E (see last chapter) and fish polyunsaturated fatty acids in the diet may have a protective role against cancer.

Although fewer cases of cancer can be linked to viruses, it has been shown that viruses are the most important class of infectious agents associated with human cancers. However, it may take years before the cancer develops, and it is almost certain that other factors such as a familial genetic disposition to cancer are also required. Also viral infections are always much more common than the incidence of cancers associated with the virus. By no means do all viral infections lead to cancer, although they may do so more commonly in those who are immunocompromised. This is clearly seen in tumours that arise after infection with the Epstein-Barr virus. This virus is regularly found in all human populations. Most people acquire it as a child and, like other *Herpes* viruses, it stays in the body of otherwise healthy people for life without any cancer risk. However, in immunocompromised patients, such as those with AIDS or long-term bone-marrow or organ transplant patients, lymphomas arise with identifiable signs of Epstein-Barr virus genes in them. There is a cancer, Burkitt's lymphoma, which is found in the malarial regions of Africa and New Guinea that is associated with this virus too. The virus has also been found in 30–50% of AIDS patients with this lymphoma, and in almost 50% of patients with another lymphoma called Hodgkin's disease. Other *Herpes* viruses have been found in AIDS patients with Kaposi's sarcoma. AIDS patients also get papillomavirus-related cancers, of the anus in men and of the cervix in women. The papillomaviruses are a large family of viruses that infect epithelial surfaces. Some of these cause warts, and there is now evidence that in some women cancer of the cervix and anogenital region is linked to these viruses. Not all women with the virus develop anogenital cancer, and it is not clear what other risk factors may influence this.

The Immune System and Cancer

All of the mechanisms discussed throughout this book may be brought into play against tumours, especially T cytotoxic cell action and natural killer cell killing. Natural killer cells, unlike T cytotoxic cells, do not have to rely on major histocompatibility complex (MHC) antigens on target cells to effect cell lysis or destruction. In addition, some T cytotoxic cells can bind to the target via antibody, thus obviating the need for MHC antigens. Tumour necrosis factor, as its name implies, is an especially valuable cytokine, and the interferons as well as interleukin-2 are used in the clinic for cancer control. The earlier names for interleukin-2 were secondary cytotoxic T lymphocyte factor and killer-cell helper factor. Both indicate the nature of the action of this cytokine.

It has been suggested that one of the main advantages of having immune system cells that patrol the body for pathogens is that they can, by using the same mechanisms, recognise cells that might become cancerous and cause them to be deleted. This is called the immune surveillance theory. In the early days of learning to transplant organs, relatively high levels of immunosuppressant drugs were used in order to stop the transplant being rejected. It was subsequently found that these transplant patients developed various cancers. By killing off immune system cells, the body could not spot and deal with cancerous cells when there were relatively few of them. It is quite possible that the long latency in the development of cancers is related to the ability of a healthy individual to eliminate small numbers of cancerous cells. In the early stages of tumour development, the numbers of cells would be so small that it is very difficult to show that this occurs. Although one tends to think of cancerous cells as those that can escape being recognised and killed, this is not entirely so. Tumour cells often possess a different pattern of surface antigens, and immunity can be raised to these antigens. However, it is not yet clear how easily these antigens can be used by the body to recognise and kill tumour cells. Is the increased incidence of cancers in older people directly related to the decline in their immune system with age, or related to the greater time available for cells to become dysregulated and cancers to develop?

Once well established, tumours seem to be hard to eliminate by natural means. This may be because they can escape from being recognised. A feature of many cancers is that they have high levels of mutation that produce aberrant cells. Cancerous or transformed cells with damaged DNA change the expression of receptors on their surface. The normal surface molecules on cancer cells become reduced in number or grouped in an abnormal manner. They are thus more difficult for immune system cells to detect. It has been found that T cells fail to identify and react with many tumour cells because there are not many MHC antigens and adhesion molecules on the cancerous cells, or the cancers produce inhibitor cytokines to stop immune reactions. The markers that are present can be classified as tumour-associated and tumour-specific antigens. Many of the tumour-associated antigens occur in foetal tissues or on normal cells, so even if the immune system could recognise them, these molecules cannot be used to eliminate cancerous cells. Tumour-associated antigens are associated with particular tumour types, for example CA125 occurs on ovarian cancers and some other epithelial cell malignancies. However, it also occurs on the epithelial cells lining the healthy cervix and the uterus or womb, and in infectious diseases at these sites. Thus high levels of this antigen in the blood may not always indicate tumours.

Tumour-specific antigens are not found on normal tissues, and they are unique for different tumours. These antigens may derive from infecting viruses or cellular breakdown products. These molecules can be taken up by antigen-presenting cells and so can be recognised by T cytotoxic cells and cell killing can occur. Cervical cancer might be controlled in this manner since there are two proteins, E6 and E7, that are only associated with this type of cancer. Indeed, relatively high numbers of T cells have been found in some patients with cervical cancer. Some new strategies for controlling cancer are being developed along these lines. One molecule called B7, when inserted into tumours, activates the immune system by signalling to T cells to attack and destroy the cell. However, the MHC antigens are reduced in number in many cancers, so the T cells still cannot make a good immune response to the presence of the antigen. Another factor that can make recognition difficult for immune cells is that many cancers have fewer adhesion molecules on their surfaces, so again the correct interactions cannot be made with cells that have the ability to kill them.

Autoimmunity, Antibodies and Altered Self

Autoimmune diseases are those in which the body's own immune system attacks itself and destroys tissues instead of foreign invaders. Although the mechanisms behind both the cause and the nature of the diseases are completely different, autoimmune diseases, like cancers, present the body with a struggle for survival. Like cancers, there is often an inherited predisposition to develop these conditions, but why they appear in some and not other genetically disposed people is an enigma. Environmental factors have been implicated, and causative links with viral and other infections are slowly being recognised. Again, as in cancers there may be a trigger that sets off or exacerbates an already existing potentially harmful situation in the body that, until then, had been kept under control. We are a long way off understanding how and why autoimmune diseases occur.

Some autoimmune diseases primarily occur in young children, others in middle age and yet others quite late in life. Some 5% of all Europeans and North Americans suffer autoimmune diseases, and two-thirds of these are women. It has been suggested that the greater incidence of autoimmune disease in women derives from the female hormones that cause interferon-γ to be produced. This, as we saw in Chapter 7, is involved in inducing autoimmune diseases.

There are two major types of autoimmune disease, organ-specific and systemic, and several other conditions that may be autoimmune, but no one has yet clearly proved them to be so. One organ-specific autoimmune disease is insulin-dependent type-1 diabetes that attacks the insulin-making cells of the pancreas, and Lupus erythematosus is a systemic autoimmune disease in which many different cells or cellular components throughout the body are damaged. From studies aimed at trying to understand the causes of each type of autoimmunity, it seems that in organ-specific diseases tolerance is lost during T cell development. This maybe because self-antigen presentation to developing thymocytes was defective, or became defective because of some other fault. It is also possible that such antigen was not taken to the thymus at all, so developing thymocytes were never selected to be tolerant to self antigens. Also if there are faults in the development of T cells and/or accessory cells secreting cytokines, then any antigen might not be presented properly. In the case of the systemic autoimmune diseases, the story is so much more complex that many more suggestions are still being considered. It would appear that tolerance must be lost by a larger number of different T cells since there are so many more antigens involved. Of course this could also be a fault at the level of T cell development, especially if the processes of removing autoreactive cells by apoptosis are disturbed.

Another mechanism proposed to account for the development of autoimmunities is called antigenic mimicry. Sometimes a pathogen has an antigen on it that closely resembles an antigenic structure normally present on cells of the body. If by chance an immune reaction is raised to this invader's antigen, then the antibodies produced by B cells might be able to react to the self antigens. Then the body has dangerous autoantibodies circulating and an autoimmune disease could be started. Fortunately the T cell receptor binding complex makes it difficult for such accidents to happen.

However, such a mechanism could explain the generation of autoimmune conditions that arise after illnesses, hormonal imbalances or severe stress, any of which could alter the microenvironments in which the cells of the body's immune system are being generated.

Autoimmunity might also arise during the fight against other non-autoimmune diseases or infections. As an immune response gets under way, cytokines are secreted that push the reactions to cell-mediated or humoral responses. These are the T helper 1 and T helper 2 cell functions discussed previously (mainly in Chapter 6). Normally this switch aids the immune response, but the factors released can cause tissue damage perhaps due to the action of cytotoxic immune cells. Then self-antigens could leak into the blood. Such a situation might allow an autoimmune response to be started against these low levels of circulating tissue antigens. This change in local microenvironment is so central to the development of autoimmune diseases, that methods of counteracting the harmful effects of the cytokines are being devised as realistic therapies for some autoimmune diseases such as multiple sclerosis, where interferon-β is being used to help prevent relapses of the disease (see Chapter 10).

10

The New Millennium

Future travels Visitors to a new environment are advised to protect themselves against unfamiliar pathogens. How wise is one to ignore such health recommendations? Why is it that the local inhabitants in these countries do not take the same precautions as visitors? Every individual, whilst growing up, is exposed to most of the pathogens of his/her local environment. We have already seen that mother's milk gives some protection to the newborn whilst the baby's immune system is developing. Then a baby born to a family that has stayed in one environment for several generations may inherit genes that have been selected for by their ability to give protection against locally occurring pathogens. In modern society though, there is so much more movement between countries than there was during the periods of human evolution, that genetic recombination is probably of less importance in practice than in theory. However, most children in western societies will be exposed to common colds and other common infections during childhood, and will adapt their immune system to cope with repeat challenges. Even more serious pathogens such as the tuberculosis bacillus are present in western society and many adults have been exposed to it, and gained some immunity. Thus the immune system is learning all the time which pathogens are around. Flying away for holidays or work to a completely new location means that suddenly the body is faced with new challenges it has not met before. Perhaps a healthy adult may stand a good chance of fighting off a mild infection, but the stress of travelling, disturbed sleep patterns and often a time-change mean that the body is less able to cope with an acute immune challenge. Then there is the problem of transporting an infection back to your own country if the body has not eliminated it in the time available. Small babies and the elderly back home may be very vulnerable, and unable to effectively fight such new pathogens. The precaution of being vaccinated and immunised is good for the individual and society in general. Many diseases can be contained if all the population is protected by vaccination or immunisation programmes. Unfortunately, these programmes are costly and if there is any requirement for follow-up treatments in poorer or mobile populations, there will be a large number of people who are inadequately protected. Thus it is often more effective in poorer populations to concentrate on improvements to hygiene, life style and nutrition rather than investing in a widespread immunisation programme.

Most of the 1000 million cases of parasitic disease world-wide occur in developing countries where the majority of local inhabitants in an area may be infected. Many die young, and those that survive do so with immense suffering and often at a huge cost to the community in terms of lost working hours. Also, even if an individual living in, for example, a malaria area appears clinically immune, the parasites still exist within the body and can be a reservoir of infection for others. If people in at-risk areas could afford to do so, and understood the benefits, many 'locals' would take the same precautions against disease that comparatively wealthy visitors are encouraged to take.

The Appearance of New Diseases

During the 1980s, the world was shocked by the appearance and rapid spread of an entirely new disease – acquired immune deficiency syndrome or AIDS. When this happens there are many questions about the origin of the disease that arise, since people are scared that other new diseases could also suddenly appear. Sporadic AIDS cases may have been in Africa for as long as a century, but it is thought that the disease was not around much before that. Something must have happened to an already existing disease to change it from an unremarkable disease to a dangerous one, or to create a new disease from an older, less virulent version. An early suggestion for the evolution of AIDS was that the causative viruses (human immunodeficiency virus, HIV-1 and HIV-2) were passed to Man from monkeys. The monkey virus, simian immunodeficiency virus (SIV), has some similarities with HIV-2 (the commonest form of AIDS virus in Africa, but not in the United States of America), but not enough to make it absolutely clear if there is a link. Another view is that AIDS is a an example of a new disease that has arisen from mutations to genes in an already existing virus. Either could be true, but what is also clear is that the disease only became a major worldwide problem when cultural habits, urbanisation, migration of peoples and long-distance travel between countries and along highways allowed the virus to spread out to new regions.

The spread of disease by Man is well known. This is what happened with Bubonic plague. Until the thirteenth century, the disease was only present in the Unnan region of Burma, where the life style of the local peoples limited the disease. With the movement of the Mongol armies across Northern Burma to the steppes of Asia, so too went the rats and their fleas that carried the plague. Once established in travellers along the caravan routes, the plague was spread to Europe and elsewhere fairly easily. Bubonic plague was coped with by the peoples of northern Burma, but to Europeans it was new and devastating. So, any threats from 'new' diseases in the future may come from the evolution of new diseases or the appearance of old diseases in new situations, when their spread is aggravated by Man's activities.

We have already seen (Chapter 6) how antigenic shift and drift cause changes in the influenza virus, creating new influenza against which people do not have antibodies. Some strain changes in other organisms also have created problems for mankind. Toxic shock syndrome, which appeared in the 1980s, is a case of a well-known organism, *Staphylococcus aureus*, developing a toxic strain that can survive in the vagina of women using tampons for the control of menstrual flow. If a tampon is left in place for a long time, then the aerobic conditions of the vagina are altered and the bacteria survive there. A change in the composition of tampons and a public education programme have almost eliminated this disease. This action has not eliminated *S. aureus* which continues to be a problem in some surgical wounds, especially as the bacteria have developed new strains that are antibiotic resistant, but the toxic shock syndrome is controllable.

Another disease that has recently become a concern in North America, Europe and Asia is Lyme disease. The first cases were associated with an arthritis-like condition and, often later, inflammation of the heart (carditis). Initially there is a ring-shaped rash, and perhaps a 'flu-like fever. The nervous system becomes infected, and the patient may suffer a range of neurological problems including a condition rather like meningitis. Arthritis and other complications may develop, and some patients are afflicted by what appears to be post-viral fatigue syndrome (see Chapter 7). Although there have been recorded outbreaks during the nineteenth century, the disease was virtually unknown in the United States of America 20 years ago. Now it is the most common arthropod-borne disease (arthropods are the group of animals that includes insects), and a ready topic of conversation amongst hikers and those enjoying out-of-doors pastimes. This is because it is easy to become infected. The disease is caused by a parasite called *Borrelia burgdorferi* that lives in ticks, *Ixodes scapularis* and *Ixodes dammini* that will bite Man. The ticks normally live on a wide range of animals and birds, but white-tailed deer are almost always tick infested. These deer became relatively uncommon after much of Northeast America was deforested in the eighteenth and nineteenth centuries so the ticks did not often bite Man. As agricultural practices changed, and re-afforestation occurred, so the deer population increased, and today it is still rising fast. Also hiking, biking and camping have become more popular, so many more people are at risk of being bitten. Although rarely life-threatening, Lyme disease is not pleasant and will be very difficult to control because of the large numbers of wild animals involved. Vaccines are being developed, and control of the carriers is being attempted in some parts of America. People can protect themselves to some extent by wearing long trousers tucked into socks, but in high infestation areas this is not enough protection, and people get bitten by the tick and infected with bacteria causing Lyme and other diseases.

The Disappearance of Old Diseases

After the development of drugs for treating infections, the increased awareness of simple hygiene regimes and the availability of vaccines for certain common but widespread diseases, it seemed possible that the international community could work together to control some of the world's worst killers, even to eradicate selected diseases. However, despite great optimism and the expenditure of much money, smallpox is the only disease to have been completely eliminated by human action. Many factors have frustrated disease control programmes. A major problem is the unequal spread of population and wealth in the world's inhabited lands. Some 90% of the world's diseases occur in the tropics, but less than 10% of the world's investment on research and development is directed to these diseases. Many disease-ridden regions are overpopulated, large numbers of the people undernourished and, in addition, too poor to purchase life-saving drugs. Even when large programmes of disease control are internationally or governmentally funded, local populations can be suspicious of the motives, uncooperative, or still too ignorant of the beneficial effects for the programmes to work.

However, Man is ever optimistic, and the World Health Organisation's last report concludes that four of the world's most important tropical diseases can be eliminated within ten years. These are leprosy, river blindness, Chagas' disease and lymphatic filariasis. Leprosy now affects almost one million people with poor living standards, who live in tropical and subtropical regions. Just over a decade ago, there were about five million cases. Due to the control measures available it is no longer a problem in Europe and North America. The disease is caused by the bacterium *Mycobacterium leprae*. As we saw in Chapter 6, these bacteria have a cell wall that cannot easily be broken down and although phagocytes such as macrophages attack the bacteria, the bacteria are not always killed, and may persist in the body for many years. The body's defence is to wall up the bacteria into granulomatous masses. There are two main forms of the disease which may occur together or separately: tuberculoid and lepromatous forms. The first, in which there are few bacteria in the granulomatous lesions, primarily affects the sensory nerves whereas the lepromatous form is characterised by skin lesions that are packed full of bacteria. These often affect the face and may become secondarily infected with other bacteria. Which form predominates reflects the strength of the cell-mediated immune response of the infected person to the bacteria. When the bacteria are not conquered (lepromatous form), the disease progresses fast. Modern approaches use a cocktail of three antibiotics to kill the bacteria whilst improving the standard of hygiene to prevent bacteria from being passed from one person to another.

River blindness is a problem in Africa and Central America. It causes blindness in over 300,000 people, and millions more suffer severe skin infections. It is caused by a round worm parasite called *Onchocerca volvulus* and it is spread by blackflies of the *Simulium* family. Many communities develop along rivers where there is adequate water for growing food and for drinking, and the blackflies live there too. As the flies bite they transmit the round worms to Man. The tiny parasites live in the skin, but after a while the body reacts by making fibrous (but itchy) nodules around them. Blindness occurs when the parasites reach the eyes by travelling along lymphatic vessels. Because river blindness makes so many regions uninhabitable, there have, in the past, been widespread insecticide programmes aimed at wiping out the biting flies. Since they live along the rivers though, the insecticides killed fish and other wildlife, and in some cases poisoned the soil needed for agriculture. Nowadays, a drug called ivermectin gives good control of the parasite in infected persons, and 11 West African countries are now free of the disease. Thus complete control may be possible. This would be of enormous advantage to the populations in these areas as it would open up the fertile river valleys for agriculture and the development of towns.

Chagas' disease is a disease of major importance in Latin America where around 45,000 deaths occur each year amongst the 18 million infected people. The protozoan parasite is one of the *Trypanosoma* family (*T. cruzi*). Another *Trypanosoma* causes sleeping sickness. Chagas' disease is caught by infecting a cut on the skin with the faeces of a bug of the Reduviidae family. Thus where the bug lives in houses and hygiene is poor, children are easily infected, often by rubbing their eyes with dirty hands. The site of entry on the skin may become inflamed, and a capsule or chagoma can be formed in the region. The parasite gets to the lymph nodes where it multiplies as a form with a long cellular extension that enables the parasite to move around in the body. When it gets into the heart, nervous system and other tissues the parasite changes its form again. Current treatments can reduce the duration of acute symptoms and thereby prolong life, but cannot eradicate the parasite once it is established in the body. Practical methods of control such as plastering walls that harbour bugs and using pesticide-impregnated curtains are helpful. As a result, the parasite is being eliminated in the southern part of Latin America, so that again total control does seem possible.

Lymphatic filariasis has been called the world's most neglected serious disease. It has been estimated that 120 million people in 73 mainly tropical and subtropical countries are infected with one of the tiny parasites, mainly *Wuchereria bancrofti*, that cause these diseases. The parasites are round worms as in Chagas' disease, but they are carried by mosquitoes of predominantly one genus called *Culex*. Once inside the human body, the adult parasites, which are about seven to eight centimetres long, get into the lymphatic vessels of the body. Then female parasites liberate tiny microfilaria (about 0.3 millimetres long) into the blood at certain characteristic times of day. Some species liberate them at night, some during the day and others liberate them all the time. Most health problems arise because the parasites cause a low-grade infection, inflammation and blockages of the lymph flow. Then the tissues around become thickened and scarred. When this is severe, swellings of the legs (elephantiasis), breast or genitalia may occur, and kidney damage is also common.

In most infected populations school children seem to have the heaviest parasite burdens. This maybe because they are more lax in their hygiene and perhaps because their immunity has not had time to deal effectively with the parasites. Thus it is often cheaper and more effective to try to treat heavily infected groups such as children rather than the whole population. When the general population is treated though, there are medicated salts that are effective and changes in simple hygiene such as encouraging people to use latrines are a help. New drugs are just about to be tested in India on 40 million at-risk people. The results of this should indicate whether eradication of lymphatic filariasis is an achievable goal.

Perennial Problems

Some diseases are much more of a problem to deal with. Malaria, which was described in detail in Chapter 6, is one of these. An enormous number of people are affected, some 270 million, living predominantly in Africa, South

East Asia and other tropical and subtropical regions. One in a hundred infected people will die of malaria. Over this widespread area, there are a number of different forms of malaria, several different mosquitoes that carry the disease, immune and non-immune sufferers, different levels of exposure to the disease, and nowadays an ever increasing population of travellers moving from country to country. Even if one aspect of malaria is tackled, the parasite itself has a variety of means by which it can alter its infectivity, drug resistance and survival in the hosts.

Many completely different factors influence the impact of malaria on humans. Some forms cause more disease than others. The *Plasmodium vivax* form of malaria only infects developing erythrocytes, whereas *Plasmodium falciparum* infects all erythrocytes and causes the greatest health problems. Not all people in a malaria area are at the same risk of becoming infected. West Africans who lack a certain blood group named Duffy do not get infected by *P. vivax*. In addition, the parasite can confuse the human body. Not only does the immune system find it difficult to start an effective immune response but, if there is a response, then the parasites can use different genes in their make-up to become polymorphic. This means that they are able to change and so alter their antigenicity. A new immune response then has to be initiated to overcome the parasite. It also means that the development of effective vaccines is difficult.

In the 1970s it was shown that, despite these difficulties, malarial vaccines could be made. A few human volunteers were injected with irradiated sporozoites, and did become resistant to this stage in the life cycle of the parasite. However, very large numbers of sporozoites would be needed to raise the vaccines and, since they cannot be maintained in the laboratory, this vaccine was not a practical answer to the control of malaria. Nowadays, the antigenic molecules on the malarial parasites are made synthetically, or by sophisticated genetic recombination procedures. A major problem is to know which antigenic molecule of the parasite to make the vaccine against. There are so many stages of parasites and so many possible variations to antigens that parasites can make that no one useful vaccine exists at present. Indeed it looks as though an effective vaccine might have to contain a number of antigenic molecules from different developmental stages of *Plasmodium* to enable the body to make the right antibodies to control the parasite. The problems are so great that, to date, there are no vaccines registered in the United States of America for use against malaria parasites. This is an important area where progress is urgently needed.

As a result, most efforts to control malaria are aimed at controlling mosquitoes, and preventing infections. Controlling mosquitoes can work. When malaria was introduced by accident into Brazil in 1938, it caused over 14,000 deaths in eight months. An active campaign to eradicate the mosquito carrying malaria in Brazil, *Anopheles gambiae*, saved the situation. Earlier antimosquito campaigns across the world used insecticides, but mosquitoes can escape being killed by shifting their patterns of feeding (and therefore biting) and behaviour as well as by becoming resistant to the chemicals used.

Now that genes can be studied in the laboratory, there is currently a million dollar effort to map mosquito genes in laboratories in Australia, the UK and the United States of America. But the life cycle of *Plasmodium* is so complex that genetic mapping is not an easy task. One particular aim is to identify parasite-inhibiting genes in the mosquitoes. If they can be identified then mosquitoes can be made with these genes in them. Once released in the wild, they would breed with the natural population and reduce the numbers of mosquitoes able to carry the infection. This approach would be most effective in some species of mosquitoes such as the West African populations of *Anopheles gambiae* and *A. arabiensis* because they do not naturally have as much genetic variation as other species. Fortunately *A. funestus*, which is not suitable for this treatment, is more susceptible to control by insecticides, so perhaps a combination of both methods of control would give the best results.

In the meantime, people living in infected areas can take simple precautions such as using insecticide-impregnated bed-nets to reduce being bitten at night. This is important since it has been estimated that in high-malaria-risk areas, a person will be bitten up to 300 times a year by infected mosquitoes. The precautions that travellers can take are simple and effective (see Chapter 3 where they are described in detail). If medicines are used they must be used before, during and after a visit to endemic malaria areas to prevent malaria from developing when one gets home.

Vaccines in the Future

Whilst inoculation is a fairly general term meaning the injection of a substance into the body, the term vaccination has a specific meaning. Vaccinations present the body with a dead or inactive fragment containing the same antigenic recognition molecules as the live pathogen. Thus vaccination allows the body to ‘see’ the pathogen’s antigen, and to react against it in a controlled manner without the risk of an infection getting out of hand. The body makes a normal but attenuated response so that when the real infection is met later, the body is already primed to react because it has developed memory cells. A brisk effective immune response occurs. Why then are there not effective vaccines for all pathogens? We have already seen earlier in this chapter and in Chapter 6 how the malaria parasites and the influenza virus can change their antigenic molecules to keep one step ahead of the antibodies the body can make. Most pathogens, especially parasites, can do this. Thus all the time and money spent over many years in developing a pure, safe and effective vaccine can be sabotaged very quickly by small changes in the pathogen. Some good vaccines do exist, but many more need to be developed. The vaccines most urgently needed throughout the world are to treat AIDS, malaria, tuberculosis, infections with respiratory syncytial virus and pneumococcal disease. Also it might be possible to make vaccines for cancers that are linked to viruses, as in the case of Epstein-Barr virus and Burkitt's lymphoma in Africa or for nasal cancer in China.

The Children's Vaccine Initiative was established in 1990 with the goal of immunising all of the world's children with existing vaccines and developing new and improved vaccines by the year 2000 – the New Millennium. The ideal vaccines are safe, heat-stable and effective when given orally by only one or perhaps a few doses early in life. Such principles are now the basis for all vaccine development. Since children are the most important group to vaccinate on a programme basis, the administration of several vaccines is made as simple as possible by giving several together, or by combining them in one preparation. This reduces the number of visits to the clinic, thereby reduces costs, and lessens the misery for the recipient. However, the combination must give a similar protection to that obtained by each one alone, the vaccines must not interact, or alter the body's natural immune response. Simple statements, yet difficult to check. Indeed, many vaccination regimes used today were devised empirically over the years from the experience of the medical profession, and need to be tested out more formally.

Commonly used composite vaccines are the combined diphtheria, tetanus toxoids and pertussis vaccines (DTP), as well as the trivalent oral polio vaccines. Fairly recently, it has been shown that a vaccine for a common bacterial pathogen, *Haemophilus influenzae* type b, that can cause meningitis can be combined with DTP. This would be a great advantage because before the vaccine was used there were about 20,000 *Haemophilus* infections annually in the United States of America, of which 12,000 caused meningitis. Once introduced the number of cases dropped to around 1250.

This was partly because the vaccinations give protection from infection, but also because the disease was not spread so widely. Originally measles vaccines were given separately, but now they are usually combined with mumps and rubella vaccines. Vaccines against smallpox can be given with those raised against yellow fever or measles. A hepatitis B vaccine has recently been introduced for infants, and this is gaining use when combined with vaccinations for diphtheria and tetanus toxoids. Another type of combined vaccine is that used against pneumococcal bacterial disease. This contains 23 of the 83 known pneumococcal types and protects against 85–90% of invasive pneumococcal disease.

Some vaccines have been developed from live viruses, and these are the ones that can interfere with others. When the trivalent oral polio vaccines are given together, the three types compete against each other. Antibody to all three generally occurs after the second dose, but to obtain reliable 90–100% responses to all three viruses, three doses of vaccine are given.

When vaccines were first developed they were raised to whole killed or live pathogens. Nowadays, synthetic fragments of antigens are generally used as they are safer and more specific. However, the fragments are often not so antigenic after injection. Thus there is now a tendency to add in factors that enhance or directly act through natural immune responses. Vaccines are being designed to enhance T cell help, either through major histocompatibility complex (MHC) components or through factors such as cytokines. Non-specific enhancers, called adjuvants, added to vaccines can direct the immune response towards the expression of specific T helper cell types (T helper 1 or 2 subtypes) and thereby influence the course of the natural response.

With greater understanding of how the immune system works has come the development of ways for enhancing the interactions between antigens used in vaccinations and immune cells. Oral antigens can be protected from destruction by the acids of the stomach, so that they can get down to the Peyer's patches in the small intestine. From here they can be rapidly taken into the body, and cause an immune reaction. Then, antigenic fragments can be designed to make a really good and effective fit with natural T and B cells. Another approach is to alter viruses in a controlled manner so that they can carry antigens into a person and make antibodies inside them. The information for this in the viral gene does not get into the person's chromosomes, but stays in the virus. The new antibody is processed by the immune system as normal, exactly as if it was produced by the body's own B cells. These forms of vaccines are called nucleic acid vaccines. These could be useful where one sort of immune response is not enough to control the pathogen, as in AIDS. One of these types of vaccine was on trial in 1995, but there are no reports yet of the success of this approach of boosting immune responses.

Another approach to developing new vaccines has also arisen from the enormous strides made recently in our understanding of genes and their molecular biology. The genes from a pathogen can be isolated and incorporated into a non-disease-forming carrier. The vaccinia virus seems to accept large amounts of genetic material without any detrimental effect on its life. Thus efforts are being made to add foreign antigen to this virus, with genes for producing cytokines that direct an immune response. Apparently a rabies vaccine has been developed in this way, and it is now being tested. The attenuated Sabin poliovirus, which has been used in Man for vaccination against polio, can also be made to carry the genes for foreign antigens. There are several variations on these techniques and many look exciting. However, the cost to the industry is high, and developers need to be assured that they can recoup their costs. Also, because of the interference with genetic material, the controls on the use of these vaccines in trials are very stringent, so it will takes many years before any can be used widely.

Manipulating the Immune System

Since disease is seen as bad, most early methods of controlling disease were directed towards switching off the immune response. Just as effective, but less explored, is the stimulation or modulation of specific immune cells in situations where the immune system is weakened by disease, or by anti-cancer therapies. The new approaches to controlling HIV infection, auto-immunities and cancers will be considered below, but before that it is worth drawing attention to some more general therapeutic approaches that use factors from the immune system to modulate, or in some cases strengthen, the immune system. To date, the major use of many is for anticancer therapy, but their central role in many immune responses makes them potential candidates for the control of many other diseases.

In addition to time-honoured traditional herbal medicines, which are difficult to define, there are three main types of immunotherapeutic agents that can act broadly to improve immune reactions. These are substances derived from bacteria and fungi, natural chemicals from the lymphoid organs, and chemically defined substances that have been shown to be important in normal immune responses. The first group include natural substances that were noted to have an effect on the immune system. One example is the bacterium *Mycobacterium bovis* that was originally used for immunisations against tuberculosis, but now is used to stimulate the immune system in cancer treatments. The muramyl peptides were originally derived from the coats around bacteria, but now they are being made artificially to produce purer substances that are not at risk of having any contaminants from natural sources. These also aid recovery following anticancer treatments, since they stimulate the bone marrow macrophages to secrete factors causing more granulocytes to be made. Thus the body's recovery from chemotherapy is improved.

There are now several natural hormone-like substances from lymphoid organs, two of which are chemically defined, which can now be reclassified as members of the third group of immunostimulants. These are thymopentin and thymostimulin which, with other less well described factors, are at the stage of being used in clinical trials in Europe as a means of improving the immune responses of cancer sufferers to other infections. Trials have suggested that these substances are best used just before or with conventional chemotherapy treatments since they appear to enhance the proliferation of T cells and increase cytokine production. In this way they boost the body's own natural defences against infections, and improve the quality of life at a difficult time. They seem to have a similar action when used with azidothymidine (AZT or zidovudine) to help AIDS patients fight opportunistic infections. In general there are few side-effects, and these natural immunoenhancers are being increasingly used in addition to more orthodox treatments. Many other licensed immunostimulants also stimulate macrophages, natural killer cells and eosinophils into action. They are thus most useful in the treatment of infections, burns and cancer patients, when the condition is being treated by other means. On their own they do not seem to be very powerful.

The chemically defined substances are most promising, and they are the largest group of factors used in immunotherapy. Most of these are being evaluated for use with AIDS and cancer patients. Many factors act in a very general manner, and give side-effects of fever and general malaise as normally seen in infections. The interferons were originally identified in cells infected with viruses, but it is now known that lymphocytes also produce interferons during an immune response. They act by producing a number of enzymes that all lead to the inactivation of viral infections. Some of the first immunomodulators to be used against viral infections and cancers caused interferon to be produced by the body, and many of these substances also increased macrophage and natural killer cell activity. However they were also toxic so they are now replaced by interferon-γ produced synthetically. Although potentially very useful, their value is limited by the side-effects of fever, fatigue, headaches, weakness, anaemia and other problems. Thus these substance have not been exploited as much as expected when they were first discovered.

The release into the body of another natural anticancer factor, tumour necrosis factor, can be stimulated by a range of cytokines and growth factors including interferon-γ. In this context interferon-γ has been approved by the Food and Drug Administration in the United States of America for the treatment of certain blood cancers. Its presence in the blood enhances the number of receptors on target cells for tumour necrosis factor, so cell killing is better. The growth factors of the colony stimulating factor family are now being tested in those with cancer who have received routine chemotherapy or radiotherapy. Since these treatments reduce the numbers leukocytes or granulocytes, the use of growth factors is expected to improve cell numbers.

Some diseases and conditions can be attacked because the immune response to them is more or less specific. The newer approaches to allergies is a good example of these types of approach. The complexities of the antigens causing some allergies make them difficult to control by the development of new vaccines. For example, a simple allergen such as Bet v 1, which is the major pollen allergen of European birch trees, occurs in at least 12 forms of the same molecule. Although it is possible that most people allergic to this type of pollen are reacting to a few similar forms, the degree of variation between people, between species of birch and the degree of sensitivity to allergens (atopy) makes the raising and use of antibodies to such allergens a major task. It also means that someone allergic to birch allergens, for example, will have T cells inside them that are reacting to lots of different forms of pollen allergen. In order to stop an allergic reaction, each separate type of reactive T cell would have to be prevented from reacting. Making drugs that do this for different people and for a range of different allergens is a formidable task. Thus ways are being explored of manipulating the immune response or of stopping the reactions to the allergens.

In Chapter 3, the importance of allergens binding to the immunoglobulin IgE on mast cells and setting off an allergic attack was described. A key player in this story is histamine since it is released by mast cells activated by allergens. If it were possible to inactive the mast cells, which produce histamine, then the allergic reaction would not be set off. This can be done by linking allergens to a mast cell inactivator called monomethoxy polyethylene glycol. Another treatment is to give allergic people specially made anti-IgE antibodies that bind to the receptors where the allergens would bind. Thus when allergens cause new IgE to be produced, it cannot occupy the receptors, and cannot start a strong reaction to the allergens. Finally a 'smart' vaccine, to be given by one injection, has recently been developed in the UK that blocks the chemical processes leading to allergies and therefore is effective against all allergies. Although extensive clinical trials are required before such a vaccine could be in general use, the prospect is hopeful for the 15 million people in the UK who suffer allergies from hay-fever, dust mites, etc.

Assisting the Fight against AIDS

Because the progress in making good vaccines against HIV infections is so difficult and slow, several ways of manipulating immune response are being developed. Many involve using the CD8 T cells since these are the prime cells that kill virus-infected cells. An infected person's CD8 T cells can be removed from the body, and the virus removed from them before they are stimulate to multiply, after which they are they are then replaced back into the patient. This gives the body a better chance of fighting the virus. Another way is to artificially create unusual T cells that will bind to HIV-infected cells. Both approaches should result in better cell-mediated killing of infected cells. However, should it be possible to protect the cleaned-up cells against HIV infection and then amplify and restore the cells, then the body might be protected against the virus. This approach is being developed, but because there are no really good animal models of HIV infection most work has to be done on cells grown in culture. It then requires many tests to be carried out before it is safe to try out any treatments in humans. This means that it will be a long time before these studies are used in Man, probably well into the next millennium.

As with the development of new vaccines, it is now possible to manipulate the genetic material inside infected cells. One current idea for controlling HIV centres around inserting a useful gene into target cells to render them resistant to the virus multiplying. By giving the patient such protected cells, the spread of the virus could be limited, and the progression of the disease slowed. The most obvious cell to choose for this is the CD4 cell, which is the major target destroyed by the virus in those with AIDS. Furthermore if the infected person's CD4 cells can be taken out, and viral replication suppressed by drug action, the cells can also be given genes for suicide molecules (see below). Such treated cells are allowed to increase in number by division before being reintroduced into the patient. The HIV-positive patient now has a population of CD4 cells that will get infected, but they will also be forced to 'commit suicide', resulting in the destruction of the virus causing AIDS.

However, it would be more effective if the T precursor stem cells in the bone marrow could be protected before they are infected. Several ways of achieving this are currently being explored. One involves nucleic acid molecules that either compete with or cut out (thereby inactivate) viral ribonucleic acid (RNA) that binds essential HIV regulatory proteins. The problems here are that effective molecules could also interfere with the normal cell division processes and cause havoc in the body. However, by judiciously choosing the RNA sequences used, some success has already been achieved. It might also be possible to make the cells cut up HIV RNA, or to protect uninfected cells with packets of cutting enzymes (ribozymes) that lead to the creation of smaller non-infective viral particles to give true protection against HIV.

A second strategy involves creating, by genetic manipulation, intracellular protein molecules that interfere with normal viral functions. Mutant proteins can be made in the laboratory that are like proteins needed by HIV for the virus to replicate. These mutants compete with the virus's own proteins and effectively suppress HIV growth. A variant of this approach is to make antibodies that sequester viral proteins in inappropriate places within the infected cell. One such antibody captures a protein that is made early in the viral replication cycle thereby stopping viral spread and limiting the initial stages of infection.

A third type of genetic manipulation involves inducing toxic molecules that kill host cells when they are first infected. These are called suicide molecules. One type of suicide molecule causes the selective killing of cells only when it is with drugs such as acyclovir or ganciclovir, which are antiviral agents. The suicide molecule converts the drug to a potent inhibitor of host cell DNA and the cell dies without the virus being replicated. Since these molecules are very toxic, stringent tests must ensure that no accidental cell killing can occur.

A major consideration for all three strategies is how to get the genes into target cells. Genetic modifications of cells have been performed outside the body, followed by reinfusion into patients, but safer carriers that cannot cause disease are being tested. These techniques, involving modifying genes, perhaps have the best potential since they can be made to be very specific, but in practice they are a long way from being available as realistic treatments for HIV infections.

Attacking Autoimmunities

Cyclosporin A and FK506, which are used to prevent rejection of transplanted organs, both act specifically against T cells. They act on the cell cycle cyclins (Chapter 9) to stop T cells from being activated by antigen and so prevent interleukin production. Their precise action has enabled them also to be used for controlling some autoimmune diseases, especially the T cell-mediated ones such as insulin-dependent diabetes mellitus and psoriasis. On the other hand they are not as effective for the antibody-mediated autoimmune diseases such as systemic lupus erythematosus and the skin condition of pemphigus. For these diseases, cyclophosphamide is better. However, the long-term use of these drugs is not desirable since side-effects or drug resistance can develop. Indeed, sometimes the use of drugs is delayed as long as possible in young people with autoimmune disease, to minimise long-term complications.

More recent approaches for some of the T cell-mediated autoimmune diseases is to use antibodies made specifically against specified T cells (mainly CD4 and CD3) so that they will be reduced in number and less aggressive. In some clinical trials results have been very good, so it is probable that future developments will be exploiting these ways of disease control. Nevertheless, there are problems in getting the antibodies. Originally antibodies were made by injecting antigen into animals such as rodents, sheep or rabbits. This starts off an immune response, and results in the animal making lots of antibodies against the injected antigen. Antibodies made in this way were harvested and used in the first clinical trials in humans. However, the human immune system recognised the antibodies as having come from another animal, and this triggered unwanted reactions. The patients became severely ill. Thus, it has become essential to find other means of making antibodies. Antibodies made in cells grown in the laboratory are not very successful for human use, so then methods were developed to make mouse antibodies in human cells. By identifying and transplanting antigen-binding sites from rodent to human cells, it is possible to make 'humanised' antibodies against which the human body does not react so adversely. This type of antibody has already been used in the clinic, and results so far indicate that they may be useful in counteracting disease. A variation on this theme is to get bacteria to make active fragments of antibodies, and to link them to activators that might trigger T cytotoxic cells, for example, to kill cells.

Another method of autoimmune disease control arises from studies indicating that a basic factor in the appearance of autoimmunities is due to some T helper 1 cells becoming autoreactive or aggressive. This adverse reaction would normally be controlled by T helper 2 cells. Thus giving the interleukins that are associated with T helper 2 cell action, or reducing the harmful interleukins produced by T helper 1 cells does seem to control some autoimmune diseases. Also injecting specific antibodies against some factors such as interferon-γ from T helper 1 cells also helps to skew the subset composition away from the damaging effects of T helper 1 cells towards the regulatory effects of the T helper 2 cells.

Naive T cells, when they leave the thymus, are not yet committed to either T helper 1 or 2 subtype, and the decision is made when the naive cell meets antigen. This offers another chance of altering T cell subsets in favour of T helper 2 cells. Studies of animals though have shown that, although the principles are correct, it is difficult to achieve this effect in practice. A slightly different approach to manipulating T cell subsets is based on animal observations that showed that the production of either T helper 1 or 2 subsets depends on how antigen is encountered by the naive CD4 cells at the time of the challenge to the immune system. Injected antigens stimulate both T helper 1 and 2 cells whereas antigen given orally elicits a strong T helper 2 response and seems to protect against active disease. Thus a combination of the route of administration plus the use of selected cytokines could favour one type of T helper subset over the other. In addition to manipulating the CD4 subset of T cells, it is also possible to target CD8 T cytotoxic or suppressor cells and alter the regulation of immune reactions. All of these approaches have been tried in animals but, as yet, none is used routinely for autoimmune disease control. It may not be long though before the techniques are perfected.

One treatment emanating from the normal immune response that has been used in the clinic is the administration of interferons. This arose as a possibility when it was found that cells from those with multiple sclerosis were deficient in the production of interferons. These cytokines, as discussed earlier in considering immunomodulators, are potent antiviral agents that increase the cytotoxicity of natural killer cells and the phagocytic properties of macrophages. Interferons can be made in cell cultures programmed with interferon genes (recombinant interferons), so that they can be produced in large quantities and high purity. Clinical trials have indicated their value in the treatment of multiple sclerosis, and advice has been published for their use in the treatment of 18–50 year olds who have a pattern of relapsing and remitting disease but are still able to walk. There could be dangers to pregnant mothers of inducing abortion, and interferons could pass to the baby through breast milk. The use of other 'natural' factors could become increasingly common now that many can be made by gene manipulation techniques. However, as with all new treatments, extensive clinical trials are always required before new therapies can be adopted.

Caring for Cancer Sufferers

It has been estimated that by the year 2000 there will be about ten million new cases of cancer diagnosed throughout the world. The actual numbers will be higher as these figures do not include non-melanoma skin cancers that are not reported accurately and are rarely fatal. This potential increase will occur because in the westernised countries people are living longer, fewer people die of other causes and cancers are diagnosed earlier. Many, if not most, cancers are avoidable. Prevention is a most important aspect of controlling cancers. Since 20–30% of all cancers in developed countries are related to tobacco smoking, if people stopped smoking this alone would have the greatest impact on cancer incidence and mortality figures than any other factor. Smokers are also often heavy drinkers. The effect of both smoking and drinking is additive, so a moderate alcohol intake and no smoking is an important life style that people could adopt to prevent them getting cancer. Another simple self-help remedy is to be careful when exposed to strong sunlight. Small children are especially vulnerable, and barrier cream protection is effective, so burning can be prevented.

There is still considerable debate as to the importance of dietary items as a cause of cancers, and of vitamins, fruit and vegetables for cancer prevention, so there are no clear cancer prevention policies in these areas yet. Even where strong causal links exist, expert advice is necessary to reduce human exposure to carcinogens. Controls need to be world-wide and rigidly enforced even where it is widely recognised that certain chemicals and/or their products, or practices such as using X-rays to diagnose diseases have been found to be associated with an increased risk of developing cancer. The world may be alerted to problems in the control, manufacture, dissemination or disposal of carcinogens, but problems are associated with the implementation of effective control measures. Finally, a fact that presently none can alter is that some people are born with a genetic predisposition to develop cancers. All of these factors mean that the development, diagnosis and treatment of cancers pose health-care cost burdens for all countries for a long time to come.

What does this mean for the future? Firstly 'at risk' people can be identified before cancers develop, and specific monitoring undertaken to catch cancers early, either before they develop, or after other forms of treatment have been given to eliminate tumours. Then there is a large field of 'chemopreventative' drugs and regimes that are being tested world-wide. These could reduce the harmful effects of exposure to carcinogenic agents, or prevent cancers from gaining hold. Diets are being analysed to see if differences in diet could explain some of the world-wide distribution patterns of specific cancers. Clinical trials are underway to assess dietary supplements or alterations to include antioxidants that reduce damage to DNA by carcinogens, intracellular inhibitors of metabolites that can cause damage, or protective factors such as an increased intake of fruit, fibre, vitamins and micronutrients. These trials include factors such as β-carotene and retinol which may act to inhibit tumour growth in lung and other cancers.

At present, most cancers, especially the leukaemias, are treated with chemotherapy. This is the use of drugs that generally interfere with the cell cycle. Some 30% are treated with radiation therapy which is the use of X-rays to kill tumours. In some ways radiation treatment can be likened to surgery, since it is directed at the tumour and is designed to remove all cancerous tissue at that site and any cells that have escaped to the draining lymph nodes. However, damage to other non-cancerous tissues can occur, although this is greatly reduced when radiation therapy is preceded by good scanning techniques to accurately localise the tumour. Both chemotherapy and radiation treatments are often combined with surgery. Which approach or what combination is used depends on the site of the cancer and its stage of development. The regimes for solid tumours fall into three main groups of chemotherapy – adjuvant, primary (also called neoadjuvant or induction) and palliative (concomitant) chemotherapy. Adjuvant chemotherapy is applied after surgical or radiological elimination of tumours and lymph nodes. It is given when there is a high possibility of really small metastases in a wide range of different organs. Primary chemotherapy involves a course of treatments given before surgery or radiotherapy, designed to reduce tumour size and activity. Its successful use can avoid too much damage to individual organs since the tumour shrinks before surgery. For example it may be possible to avoid the complete removal of a breast if the tumour can be reduced to a small lump that can be cut out. A combined use of surgery, chemotherapy and radiotherapy has been shown to be an excellent means of treating oesophageal and anal cancers although side-effects can be harder to control. Palliative chemotherapy is offered with the aim of either prolonging life or controlling symptoms caused by widespread disease.

Over the last ten years or so, a range of new cancer chemotherapy drugs generally based on older well tried drugs have been introduced. Almost all interfere with the cell cycle, and will therefore also affect any fast proliferating system such as the cells lining the gut, the hair follicles and blood-forming cell lines. It is the drug action on these dividing cells that causes the common side-effects associated with cancer chemotherapy: diarrhoea, vomiting, hair loss and an increased susceptibility to infection and anaemia. Should these side-effects occur then there are excellent medical treatments to control diarrhoea and vomiting, and antibacterial and antifungal drugs reduce infections resulting from the inability of the weakened immune system to fight infections. Also, before chemotherapy, growth factors or stem cell transplants may be used in specific cases where high doses of drugs that suppress bone marrow activity are to be used. Hormone treatments are also administered when the patient has hormone-sensitive cancers such as prostate or breast cancer. Many chemotherapeutic drugs used in the treatment of cancer work by inducing DNA damage in the tumour cells by a variety of mechanisms. Such damage results in the tumour cell undergoing cell death by apoptosis. However, some cancers cannot be killed in this way, either from the outset or because they develop resistance after drug treatment. The tumour cells survive and divide to form other resistant cells so the tumour continues to grow despite chemotherapy. Now completely different approaches are being developed to kill tumours.

It is clear that the immune system has cells and factors that are able to kill cells in response to invaders. So cytokines such as interferons and interleukin-2, which are normally secreted in this type of immune response, have been injected into cancer patients. The family of interferons have many actions already discussed in this chapter, but of importance here is that they slow down cell proliferation. They can give rise to quite marked side-effects such as fevers, gut problems and chronic fatigue, so their use has to be carefully monitored, but they can kill tumours. Interleukin-2 also has nasty side-effects when used at high doses. Despite this, it was approved by the US Food and Drug Administration in 1992 for use against specific advanced renal (kidney) cancer, and as such was the first of the 'natural' factors from the immune system to be registered. Interleukin-2 binds to T cells, activates them to secrete other cytokines, causes them to multiply and so upregulates immune system responses. It also acts on some T cells to create LAK, or large activated killer, cells that destroy tumour cells.

One of the major problems of cancer is that it can release cells to form metastases far from the original tumour. New therapies are being tried out now to inhibit growth of the invading edge of the tumour by limiting the formation of new blood vessels in the region. New blood vessels only normally form during wound healing, and in the ovaries and endometrium during the menstrual cycle. Once the tumour is stabilised, the release of metastatic cells is reduced. The action of cytotoxic drugs is enhanced too by using these agents, so combined therapies look quite promising for small tumours or for long-term stability of the disease.

The prospect of using newer techniques currently under development which target the cancerous cells directly without, theoretically at least, having any effect on other cells is very exciting. When monoclonal antibodies were first made outside the body, it seemed as though they would provided a means of identifying cancerous cells, and they have been used since the late 1980s for controlling cancers, but only with limited success. Now a whole range of new antibodies, called bispecific antibodies, have been developed that combine the ability to bind specifically to cancerous cells as well as having a killing agent on them.

The first problem is how to identify the tumours, especially since some tumours have heterogenous cell types in them. Many cancers appear to develop because they escape the immune system's surveillance and are composed of relatively non-antigenic cells, or cells that express tumour-specific antigens or oncogenes (see Chapter 9). People with cancer can mount an immune response though, as evidenced by the presence of IgM or IgG antibodies in their blood. Success will depend on discovering the best antigenic molecules on cancerous cells that can be exploited for the synthesis of antibodies. The bispecific antibodies used so far in clinical trials are directed to specific receptors on ovarian cancer cells, or to tumours like breast, ovarian and prostate cancers that have an oncogene product on their surface. Where a viral basis is suspected for the cancer, as in cervical cancer, antibodies to viral products may be useful too.

The killing agents of bispecific antibodies are toxins or drugs that are radiosensitive. This means that they kill when activated by X-rays. Other approaches utilise the body's own cells, especially T cytotoxic cells. These cells will home in accurately on to tumour cells but the disadvantage is that each T cell kills only one tumour cell, and many people with cancer, especially at an advanced stage, lack the normal complement of immune cells. Some protocols that require the use of the patient's own cells expand or activate the cells outside the body, then combine them with the tumour-seeking antibody before they are returned to the patient. The immune reaction that takes place involves the secretion of cytokines that, as we have seen in this book, have a wide spectrum of activities. Sometimes the patient suffers fevers and chills, but the tumours do appear to regress. In the future, patients are likely to have regular chemotherapy or radiotherapy first, then any residual cancerous cells may be eliminated with the bispecific antibodies. Other new bispecific antibodies are targeting monocytes and macrophages that get inside the tumours, create a local inflammation there and then kill the tumour cells. Others are being used in combination with cytokines such as interleukin-2, which activates natural killer cells. Some of these antibodies can work like the normal immune response and it seems possible that memory cells could also be produced to protect the body against the re-emergence of tumour cells in the body.

Cancer gene therapy has the capability of being a very precise means of targeting cancerous cells. Genes or DNA can now be introduced on carriers, called vectors, into tumour cells or normal cells, using some of the principles mentioned in the discussion of vaccine development above. Defective genes can be repaired, or genes made to interfere with the action of the rogue genes that made the cells cancerous. Genetically modified cells can perform functions that previously would not have been possible. For example, tumour cells can be given a gene that has the ability to make an enzyme which converts an otherwise non-toxic drug into a cytotoxic agent that will kill the tumour. These approaches are very new and, although they have been tested in the laboratory, they are only just beginning to be used in trials. Although their use looks promising, how effective they are in treating cancer patients is still very unclear. Obviously it is important to establish that the new gene only gets into tumour cells and cannot harm healthy cells.

The last decade or so has seen a much greater understanding of how genes control cellular events. The recognition of the fact that some people have genes that predispose them to cancers if the genes are changed or mutated has now been exploited, so that finding such a gene in a person can enable them to be monitored carefully for early signs of cancer. The association between breast cancer and *BRCA* genes was mentioned at the start of the last chapter, and a predisposition to develop colonic cancers may be related to one of three genes that interfere with repairing damaged DNA. The continuing presence in the body of cells with such mutated genes may also indicate that the tumour has not been killed, so testing for these cells can indicate the efficacy of treatment, or predict the future outcome of the disease. As more genes are identified, the possibility of assessing the risk of developing a cancer, and what the outcome is, will become more common, as will the problems of how to handle the knowledge. The results of genetic testing could influence a couples' decision to have children, their insurance cover, the ease of obtaining a house mortgage and many other events of life. People seem to want to be tested, but until this information can be linked to better cancer control, then testing must be applied judiciously.

Final Thoughts

The great breakthrough in understanding the immune system came with the identification of firstly the B cell receptor, and then later the T cell receptor. Both are unique in how their genes are rearranged to enable recognition of a diversity of antigens, so that the development and function of self and non-self identities became central dogma for immunologists. Quite recently, since starting to write this book, this has been challenged by the view that actually the body's ability to accept programmed cell death and to react against unnatural cell death is really the basis for immune responses. In view of the high degree of development of the antigenic recognition processes, and the way in which parasites and our immune system have evolved side by side, the concept of cell death being central to immune responses seems rather simplistic. However, cell death is fascinating. It is not morbid; it is life-giving by eliminating the damaged, diseased and unwanted cells. Cells need to die for our bodies to survive; it is uncontrolled cell death that is dangerous.

In the future it is probable that the greatest advances in disease control will come from the greater understanding of genes, their interactions, products and repair mechanisms. Scientists will, in the near future, have described the entire human genome. Every gene will be placed on a map, and already genetic manipulation is a reality. Even synthetic chromosomes, made outside the body, can be inserted into cells which subsequently divide.

Thus many diseases will be controlled genetically in future. So too may the genetic structures that control our own death. At present though, despite identifying the part of the chromosome that limits the ultimate number of cell divisions, no one has found the means of fundamentally increasing longevity. Whilst dying to live helps us to keep a healthy body and enjoy life, we must face the fact that we live to die! Fortunately, our children can continue life.

Glossary

Activated cell – has recognised and interacted with antigen
Adaptive immunity – sophisticated antigen-recognition immune responses
Adenoids – lymphoid tissues of nasal region
Adhesion molecules – facilitate cell contacts and interactions
Adjuvant – non-specific enhancer of a response to antigens
Affinity – strength of binding
Allergen – substance that causes an allergic reaction
Amino acids – protein building blocks
Anaphylaxis – an IgE-mediated sudden change in muscles and blood vessels
Anergy – tolerance whereby lymphocytes recognise but do not respond to antigen
Antibody – a molecule from B and plasma cells that binds to an antigen
Antigen – a molecule or part of it that binds to antibody
Antigen-presenting cells – Macrophages, dendritic cells etc. that cut, process and present antigen
Antiserum – fluid collected after blood clots, that contains different immunoglobulins to antigen
Apoptosis – programmed cell death
Autocrine – secretions that act on the cell producing them
Autoimmune diseases – arise when the body's immune cells attack the body itself
B cells – lymphocytes that become antibody producing plasma cells
Basophil – granular leukocyte of blood, like a mast cell
CD numbers – Clusters of differentiation antigens with the same function
Cell cycle – phases of growth (G_1), DNA replication (S), and time (G_2) before DNA division (M)
Cell-mediated immunity – immune reactions involving cell killing
Clones – cells with identical sets of genes on chromosomes
Colony stimulating factors (CSFs) – cytokines favouring lymphoid cell line development

Complement – blood factors that mediate opsonisation, lysis of cell membranes and inflammation
Cytokines – molecules used for cell–cell signalling, e.g. growth factors
Cytotoxic – cell killing
Degranulation – action of releasing granule contents from granulocytes
Delayed hypersensitivity – inflammation developing some time after antigen recognition
Dendritic cell – a form of antigen-presenting cell in lymphoid organs
Domain – a region of a molecule (e.g. immunoglobulin) with a coherent function
Effector cell – cell that carries out an action
Encoded – information in a gene on a chromosome
Endocytosis – process of taking particles or molecules into a cell
Eosinophil – granular leukocyte involved in allergic and parasitic reactions
Epitope – part of antigen binding to antibody
Erythrocytes – red blood cells (without nuclei)
Fc – non-variant part of immunoglobulin; binding activates complement factors
Genome – all genetic material in a cell
Genotype – all of an individual's genes (not necessarily expressed)
Germinal centres – areas of B cell proliferation in lymphoid organs
Granulocytes – neutrophils, eosinophils, basophils and mast cells
Histocompatibility genes – encode MHC (mouse) and LHA or human leukocyte antigen (human) recognition proteins
HIV – Human immunodeficiency virus (the cause of AIDS)
Humoral immunity – based on immunoglobulin/antigen recognition
Hyperthermia – body temperature higher than set point (body feels hot)
Hypothermia – body temperature lower than set point (body feels cold)
IgA – secretory immunoglobulin protecting epithelial surfaces
IgD – first produced immunoglobulin on B cell's surface
IgE – immunoglobulin produced in allergic and some parasitic reactions
IgG – main immunoglobulin of B cells (several subtypes)
IgM – Pentameric immunoglobulin, fixes complement, produced early in responses
Innate immunity – evolutionarily primitive defences, primarily phagocytosis
Inoculation – injection of substances
Interferons – cytokines enhancing antiviral immunity
Interleukins (IL) – cytokines passing signals from cell to cell

Invader – used here to mean pathogen or substance evoking an immune response
LAK – large, activated, cytotoxic granule-containing leukocyte
Ligand – linking or binding molecule
Locus – where a gene is on a chromosome
Lymph – fluid in lymphatics carrying predominantly leukocytes
Lymph nodes – nodules where immune interactions occur
M cells – form the dome of Peyer's patches
Macrophage – mature phagocytic cell in tissues (derived from a monocyte)
Mast cell – cell full of bioactive granules, important in allergic reactions
Meiosis – cell division to form gametes (allows genetic recombination)
Memory cell – produced after antibody/antigen interaction
MHC – major histocompatibility complex of genes producing recognition molecules
MHC class I antigens – surface proteins on all nucleated cells
MHC class II antigens – surface proteins on antigen-presenting cells
MHC class III antigens– molecules of mainly complement system
Mitosis – cell division without genetic recombination
Monoclonal – all of one type (identical)
Monocyte – precursor of macrophage, circulates in the blood
Myeloma – cancerous proliferation of B cells
Naive cell – not yet involved in immune responses
Neoplasm – tumour
Neutrophil – phagocytic granular leukocyte
NK cells – natural killer cells (act against viruses and tumours)
Normothermia – when body's set and normal temperatures coincide
Opsonins – blood factors that coat antigen to facilitate phagocytosis
Pathogen – any organism causing disease
Peyer's patches – lymphoid regions in small intestines of gut
Phagocytosis – natural means of ingesting particles into a cell
Phenotype – the products (result) of an individual's genetic activity (*see* genotype)
Plasma cell – a mature immunoglobulin-producing B cell
Polyclonal – derived from several different (non-identical) cells
Prostaglandins – factors affecting cell mobility and immune responses
Pyrogen – cytokine inducing a fever
Receptor – cell-surface binding molecule
Recombination – rearrangement of genes in meiosis or T and B receptor development

SLE – systemic lupus erythematosus (autoimmune disease affecting multiple sites)
Suppressor cell – CD8 T cell able to reduce or stop immune reactions
Switching – production of a different type or subtype of immunoglobulin
Synergistic – cooperation giving added value
T cells – developed from thymocytes, have surface CD3 molecules
T cytotoxic cells – CD8 T cells that kill other cells
T helper cells – CD4 T helper 1 or 2 cells that secrete cytokines to direct immune responses
Tolerance – specific unresponsiveness in immune reactions
Tonsils – generally means lymphoid tissues at back of throat
Vaccination – giving antigen to prime body for later infection with same antigen

Index

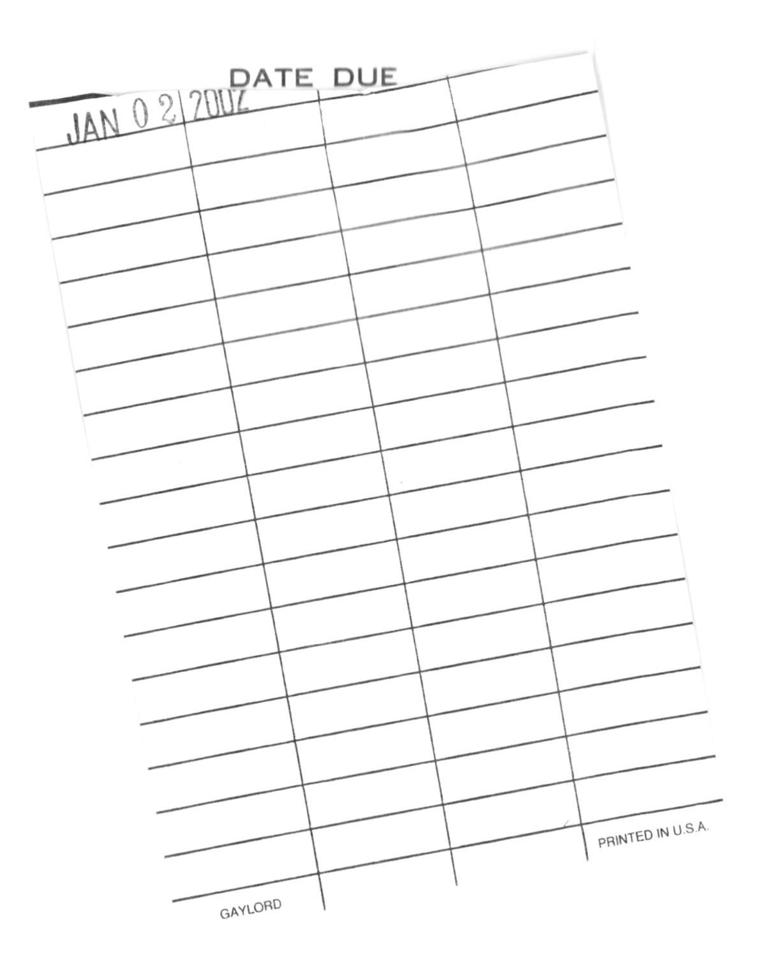
DATE DUE
JAN 0 2 2002
GAYLORD
PRINTED IN U.S.A.